OPTIMAL AUTOMATED PROCESS FAULT ANALYSIS

OPTIMAL AUTOMATED PROCESS FAULT ANALYSIS

Richard J. Fickelscherer

FALCONEER Technologies

Daniel L. Chester

FALCONNER Technologies and
Department of Computer and Information Sciences
University of Delaware

AIChE

A JOHN WILEY & SONS, INC., PUBLICATION

Library of Congress Cataloging-in-Publication Data:

Fickelscherer, Richard J.
   Optimal automated process fault analysis / by Richard J. Fickelscherer, Ph.D., P.E. co-founder,
Falconeer Technologies, LLC & Daniel L. Chester, Ph.D. co-founder, Falconeer Technologies, LLC &
associate chair, Department of Computer & Information Sciences, University of Delaware.
         pages cm
   Includes index.
   ISBN 978-1-118-37231-9 (cloth)
   1. Chemical process control–Data processing.   2. Fault location (Engineering)–Data processing.
I. Chester, Daniel L.   II. Title.
   TP155.75.F527 2013
   660′.2815–dc23

                                                                                2012023637

CONTENTS

FOREWORD

It is an honor to be asked to write the foreword to Rich and Dan's book on diagnostic reasoning for process plants. The story of the FALCON diagnostic system goes back to when I first joined MIT as a young faculty member in the early 1980s. In those days computing meant numerical computation in Fortran or C, mainframes and minicomputers ruled, and personal computers were underpowered novelties. Process control had begun a slow transition from pneumatic to digital instrumentation, but the first digital controllers were modeled unimaginatively after the PID loops they were replacing.

But Metcalfe's law was in full exponential ascent, and the world was rapidly changing, not only in terms of faster numerical methods. New ideas from artificial intelligence were flooding across the MIT campus, upending the very foundation of computing by means of an entirely new synthesis of symbolic, object-oriented, neural, and rule-based computing. Touching off a great intellectual ferment in chemical engineering, virtually every aspect of process operations was being transformed. Gauges transformed into graphical operator interfaces, fixed threshold alarms into intelligent monitoring and diagnosis, and steady-state operation into dynamic economic optimization.

The University of Delaware was one of the leaders in this exploration, especially in the area of process monitoring and fault diagnosis. Undertaking a joint project with Foxboro and DuPont in the early 1980s, Delaware spearheaded the first industrial application of an expert system, FALCON (fault analysis consultant), for online fault diagnosis of the DuPont adipic acid plant in Victoria, Texas. A key idea, expounded further in this book, was the synthesis of logical (pattern or rule) analysis and quantitative mathematical modeling.

This period of creative experimentation reached its zenith in 1995, at the First International Conference on Intelligent Systems in Process Engineering in Snowmass, Colorado. By that time, I had moved to an MIT spin-off, Gensym Corporation, developers of the G2 real-time expert system development environment. G2 was that generation's ultimate synthesis of graphical UI, structured natural language, object-oriented programming, and rule-based processing. The conference showcased innovative knowledge-based systems ranging from product design to intelligent control, optimization, and diagnostics. Flush with the success of building and deploying hundreds of expert systems to solve real industrial problems, few of us realized just how quickly another revolution, the Internet, was going to overturn everything, yet again.

Throughout these many changes, some determined individuals have persevered to bring the vision of intelligent operations closer to reality. In this book, Rich and Dan explain how they transformed the art of the diagnostic expert system into a practical and reproducible system, implemented in FALCONEER™ IV to increase the operating safety and reliability of process systems. Their system captures many lessons learned during the rapid and often convulsive change of the past 25 years. I wish them the very best.

MARK A. KRAMER, PH.D.

Winchester, Massachusetts
February 2012

PREFACE

Process fault analyzers are computer programs that can monitor process operations to identify the underlying cause(s) of operating problems. A general method for creating process fault analyzers for chemical and nuclear processing plants has been sought ever since the incorporation of computers into process control. The motivation has been the enormous potential for improving process plant operations in terms of safety and productivity. Automated process fault analysis should help process operators (1) prevent catastrophic operating disasters such as explosions, fires, meltdowns, and toxic chemical releases; (2) reduce downtime after emergency process shutdowns; (3) eliminate unnecessary process shutdowns; (4) maintain better quality control of process products; and (5) ultimately, allow both higher process efficiency and higher production levels.

A wide variety of logically viable diagnostic strategies now exist for automating process fault analysis. However, automated fault analysis is currently still not widely used within the processing industries. This is due mainly to the following limitations: (1) the prohibitively large development, verification, implementation, or maintenance costs of these programs; (2) an inability to operate a program based on a given diagnostic strategy continuously online or in realtime; and (3) an inability to model process behavior at the desired level of detail, thus leading to unreliable or highly ambiguous diagnoses. Subsequently, a method for efficient production of automated process fault analyzers is still being actively sought. It is our contention that evaluating engineering models of normal process operation with current process data is the most promising and powerful means of directly identifying underlying process operating problems. Doing so generates an unimpeachable source from which to logically infer the current state of the process being modeled. Performing this inference automatically online enables these programs to perform *intelligent supervision* of the daily operations of their associated process systems. It makes possible a fundamental understanding of a given process system's design and operation to be utilized in evaluating its current operating conditions.

The *method of minimal evidence* (**MOME**) is a model-based diagnostic strategy for developing optimal automated process fault analyzers. It was derived at the University of Delaware while developing the **FALCON** (fault analysis consultant) *system*, a real-time online process fault analyzer for a commercial-scale adipic acid plant formerly owned and operated by DuPont in Victoria, Texas. It provides a uniform framework for examining both models of normal process operation and their corresponding associated modeling assumptions that are required to build such fault analyzers. MOME can be used directly to correctly diagnose both single- and multiple-fault situations, to determine the strategic placement of process sensors to facilitate fault analysis, and to determine the shrewd division of a large process system for distributing fault analyzers.

The MOME diagnostic strategy was again demonstrated to be effective in a commercial-scale persulfate plant owned and operated by FMC in Tonawanda, New York. Versions of two *knowledge-based systems* (**KBSs**) developed using MOME have been running online at this plant since February 2001. In the current implementation, these KBSs [a.k.a. **FALCONEER**^{TM} **IV** (**FALCON** via engineering equation residuals IV)] diligently perform automated sensor validation and fault analysis of both FMC's electrolytic sodium persulfate and liquid ammonium persulfate processes in realtime. The development effort for these two FALCONEER IV applications was more than two orders of magnitude less than that required for the original FALCON system, with even better performance to date. This impressive improvement in the development and maintenance effort required was possible because FALCONEER IV contains a compiler program that automatically generates the *sensor validation and proactive fault analysis* (**SV&PFA**) *diagnostic logic* required to perform competent fault analysis directly from the underlying engineering models of normal process operation. Since the MOME diagnostic strategy is a systematic procedure, creating an algorithm based on it and then codifying that algorithm proved to be straightforward. This treatment describes both the underlying logic of MOME and the fuzzy logic algorithm based on it. It is meant to be a study guide for those who wish to develop such fault analyzers for their own process systems.

Motivations for automating process fault analysis are described in detail in Chapter 1. Our patented methodology for automating process fault analysis (MOME and its associated fuzzy logic algorithm) is then discussed in detail in Chapters 2 to 5. The logic behind model-based reasoning in general and MOME in particular is described in Chatper 2. The MOME logic for performing single- and multiple-fault diagnosis is described in Chapter 3. Also discussed in Chapter 3 are the motivations behind the creation of process fault analyzers based on MOME automatically via the SV&PFA diagnostic rule logic compiler program contained in FALCONEER^{TM} IV. The fuzzy logic algorithm automating MOME as implemented in this compiler is described in Chapter 4. In Chapter 5 the criteria for shrewdly distributing process fault analyzers throughout a large processing plant are described. Some general guidelines for the strategic placement of process sensors for directly facilitating fault diagnosis are also discussed.

Chapter 6 covers the need to augment process fault analysis with trend analysis of the various process sensor measurements and *key performance indicators* (**KPIs**) via the *virtual statistical process control* (**virtual SPC**) technique of calculating and analyzing *exponentially weighted moving averages* (**EWMAs**).

The need to first determine the current overall operating state of the process undergoing automated fault analysis in discussed in Chapter 7. Such determinations provide the proper context required for the fault analyzer to make legitimate diagnoses.

Chapter 8 summarizes the benefits derived and lessons learned when employing FALCONEER™ IV in actual process applications. A systematic procedure to follow when creating such applications is also described. The chapter concludes by summarizing the advantages of distilling the raw information contained in typical process sensor data continuously into value-added knowledge concerning the current state of process operations and having that knowledge be instantaneously available for *intelligent supervision* of those operations.

For completeness, four appendixes have been added to this treatment as background information. A number of the other various possible diagnostic strategies also used to automate process fault analysis and their limitations are reviewed briefly in Appendix A. Appendix B describes DuPont's adipic acid plant and the original automated process fault analyzer (i.e., the FALCON system) developed for it. The lessons learned from the development of this real-world fault analyzer are discussed in detail. The advantages of using the knowledge-based system paradigm for solving problems, especially those that led directly to the creation of the MOME diagnostic strategy, are also discussed. As described throughout the book, this strategy has since been codified into FALCONEER™ IV. Appendix C outlines the logic that was used by the original hand-compiled FALCONEER system to determine the current process state in FMC's electrolytic sodium persulfate plant. This logic has since been simplified, generalized, and codified in the current implementation of FALCONEER™ IV. Finally, Appendix D describes two downloadable FALCONEER™ IV demos provided to accompany this treatment.

RICHARD J. FICKELSCHERER
DANIEL L. CHESTER

ACKNOWLEDGMENTS

First, we would like to thank the other two key chief investigators at the University of Delaware on the original FALCON project, Professors Prasad S. Dhurjati and David E. Lamb, along with many student research assistants, including Oliver J. Smith IV, George M. A'zary, Larry Kramer, Dave Mooney, Lisa Laffend, Kathy Cebulka, Apperson Johnson, and Bob Varrin, Jr. We would also like to thank DuPont and its employees Duncan Rowan, Rick Taylor, John Hale, Robert Wagner, Bob Gardener, and Tim Cole, and especially our domain expert, the late Steve Matusevich. At Foxboro Inc., we would like to thank Dick Shirley, Dave Fortin, and the late Terry Rooney. Further, we thank Oliver J. Smith IV and Duncan Rowan for reviewing earlier versions of this treatment; their comments about ways to improve it were invaluable.

We also thank the FMC Corporation for its support in developing the various FALCONEER and FALCONEER™ IV KBS program applications, especially Doug Lenz, John Rovison, Charlie Lymburner, Weidong An, Don Lapham III, John Helieko, Don Stockhausen, and Jim Kaylor. As a cofounder of FALCONEER Technologies LLC, Doug Lenz strongly advocated including virtual SPC capabilities in FALCONEER™ IV; we would like to acknowledge the programming efforts of Lee Daniels required to accomplish both this and the streamlined state identification capability also currently included in FALCONEER™ IV.

Feel free to contact either of us if you would like to find out more about the FALCONEER™ IV software and our company's services. Following is the contact information:

Dr. Richard J. Fickelscherer, PE
Tonawanda, New York
falconeertech@verizon.net

Prof. Daniel Chester
Newark, Delaware
chester@cis.udel.edu

1

MOTIVATIONS FOR AUTOMATING PROCESS FAULT ANALYSIS

1.1 INTRODUCTION

Economic competition within the *chemical process industry* (CPI) has led to the construction and operation of larger, highly integrated, and more automated production plants. As a result, the primary functions performed by the process operators in these plants have changed. An unfortunate consequence of such changes is that the operators' ability to perform process fault management has been diminished. The underlying reasons for this problem and the methods currently used to counteract it are discussed here.

1.2 CPI TRENDS TO DATE

The CPI constitutes one of the largest and most important segments of the global economy. While developing into its current, relatively stable position, competition for market share among the various chemical producers has greatly intensified. This competition has, in turn, created continuously downward pressure on the market price, and hence the associated profit margin, of most commodity chemical products. Several major trends within the CPI in the operation of production plants have resulted.

Optimal Automated Process Fault Analysis, First Edition.
Richard J. Fickelscherer and Daniel L. Chester.
© 2013 John Wiley & Sons, Inc. Published 2013 by John Wiley & Sons, Inc.

One of these trends exploits the economies of scale inherent in chemical manufacturing as a means to reduce costs. This has led to the construction and operation of plants with ever-larger production capacities. While such facilities represent enormous capital investments, fixed costs per unit of production have been reduced substantially. Moreover, operating these larger plants has also reduced the direct labor costs because relatively fewer process operators are required per unit of production. As a result of this trend, most commodity chemicals are currently produced at facilities known as *world-class* plants.

Another major trend within the CPI has been the automation of the various process operations, especially process control functions. The motivation for automating process control functions is that it results in applying the best available process control strategies more accurately in a continuous, consistent, and dependable manner [1, 2]. This automation has been made possible by advances in both computer technology and process control theory. Such advances have made automated control more economically feasible, reliable, and available [1]. Process computers have also provided a significant means for dealing with the diverse and complex information required to operate a modern production plant effectively [2]. Together with advances in electronic instrumentation, these developments have led to centralized control rooms that require considerably fewer personnel to operate [1].

A third major trend designed to reduce production costs has resulted from attempts to use energy more efficiently. These have included the application of traditional conservation measures, such as adequately insulating process equipment, and various measures designed to recover and reutilize energy more effectively. The latter measures have been a direct cause of greater process system integration. This has, in turn, increased functional coupling among the various process subsystems, thereby making the operation of these subsystems highly interdependent. These interdependencies complicate operation of the overall process system, making it more difficult to start up, shut down, and control during production runs. It also opens up the possibility that a malfunction in one subsystem will cause malfunctions in other subsystems connected to them functionally.

A similar situation has resulted from the trend toward maintaining smaller inventories of raw materials and intermediate products. This complicates process operation in two ways. Since smaller buffers exist between the process subsystems, the effects of a malfunction in one subsystem can more easily migrate to other subsystems. In addition, if one subsystem is shut down for a prolonged period, it may force subsystems connected to it to be shut down. The trend toward greater process system integration and that toward limited storage facilities have a common consequence: They both make effective operation of the overall process system more critically dependent on the coordinated, faultless operation of its process subsystems.

A final trend for reducing production costs has been to maximize the availability of the plant for production. This is typically accomplished by optimally scheduling the production runs and by minimizing the effects of unexpected production disruptions. A variety of methods are in use to either eliminate or minimize the severity of unexpected production disruptions. Nonetheless, as the complexity of the plants has increased, making plants available for production has become much more difficult because the number of potential operating problems has also increased [3]. This tends to increase the frequency of unexpected production disruptions. Consequently, maximizing plant availability for efficient process operation has become more dependent on effective management of its various potential operating problems [4].

1.3 THE CHANGING ROLE OF PROCESS OPERATORS IN PLANT OPERATIONS

The process operators' main task in plant operation is to assess the process state continuously [1] and then, based on that assessment, to react appropriately. Process operators thus have three primary responsibilities [5]. The first is to monitor the performance of the various control loops to make sure that the process is operating properly. The second is to make adjustments to the process operating conditions whenever product quality or production efficiency falls outside predefined tolerance limits. The operators' third, and by far most important, responsibility is to respond properly to emergency situations: in other words, carry out effective and reliable process fault management. Such management requires that the operators detect, identify, and then implement the correct counteractions required to eliminate the process fault or faults that are causing the emergency situation. If process fault management is performed incorrectly, accidents can occur, as they have on many occasions.

The biggest change in the functions performed by process operators has been caused by the increased automation of process control. Operators now monitor and supervise process operations rather than controlling them manually. Moreover, increasingly, such functions are accomplished with interface technology designed to centralize control and information presentation [6]. As a result, their duties have become less interesting and their ability to carry out manual process control has diminished. Both situations have increased the job dissatisfaction experienced by process operators [6] and have diminished the operators' ability to perform process fault management.

A second change in the functions performed by operators in modern plants has resulted from having fewer operators present. Each operator has become responsible for a larger portion of the overall process system. This increases the risk of accidents because relatively fewer operators are available at any given time to notice the development of emergency situations or help prevent such situations from causing major accidents. In addition to the increased risk, the potential severity of accidents has also increased because larger quantities of reactive materials and energy are being processed. This makes the operators' ability to perform process fault management much more critical for ensuring safe operation of a plant.

One method used to help reduce the risk of a major accident has been the addition to the overall process control system of emergency interlock systems. Such systems are designed to shut the process down automatically during emergency situations, thereby reducing the likelihood of accidents that could threaten human and environmental well-being or damage process equipment. Emergency interlock systems therefore help ensure that a process operation is safe during emergency situations by decreasing the effects of human error [7]. Eliminating such accidents also protects the operational integrity of the process system, which in turn allows it to be restarted more quickly after emergency shutdowns.

However, the widespread use of emergency interlock systems has caused the operators' primary focus in plant operations to change from that of process safety to that of economic optimization [8]. In emergency situations, operators are now more concerned with taking the corrective actions required to keep the process system operating rather than those that will shut it down safely. They rely on the interlock system to handle emergency shutdowns, trusting that it will take over once operating conditions become too dangerous to let production continue.

A potential problem with this strategy is that to keep the process system operating, operators may take actions that counteract the symptoms of a fault situation without correcting the situation itself [9]. Such behavior by the operators may cause them inadvertently to circumvent protection of the emergency interlock system, thereby creating an emergency situation which they falsely believe to be within that protection. Another potential problem of this strategy is that the interlock system may fail, which again will create a situation in which the operators falsely believe that the process system is protected by the emergency interlock system. These potential problems can be reduced by (1) prudent design of the interlock system, (2) being certain to add sufficient redundancy to detect critically dangerous situations [7], (3) establishing a formal policy by which particular interlocks can be bypassed during process operation [10], and (4) adequate maintenance of the interlock system [11].

In summary, the automation of process control duties and of emergency process shutdowns has shifted the operators' main activities away from direct process control to that of passive process monitoring. Moreover, automation has also tended to shift their primary emphasis away from process safety to that of economic optimization. As a result of these changes, the operators' ability to perform process fault management has been reduced. Unfortunately, this reduction has occurred during a period when such management has become more critical to both the safe and economical operation of the production plants. In response, various methods have been developed to help counteract this decline in human capability with process fault management.

1.4 METHODS CURRENTLY USED TO PERFORM PROCESS FAULT MANAGEMENT

A variety of methods have been developed either to reduce the occurrence of process faults or to help operators perform process fault management more effectively when it is required. The methods currently used to reduce the occurrence of process faults include (1) designing process systems with greater operational safety in mind; (2) constructing process plants that have better quality, and therefore more reliable, process equipment; (3) implementing comprehensive programs of preventive maintenance; and (4) establishing standard operating procedures and following them strictly. The direct methods currently used to help operators perform process fault management include (1) extensive training of operators in process fault management, (2) adding alarm systems to process control systems, (3) adding emergency interlock systems to process control systems, and (4) designing better control consoles and human–machine interfaces. Each of the eight methods, along with their associated shortcomings, is discussed below.

It is useful first, though, to examine how the failures in chemical plant operations are distributed by frequency. A survey of chemical plant failures [12] in the past has shown that operational failures account for 49% of the total number of failures, while human failures account for another 32% of that total. Equipment failures account for approximately one-half of all operational failures. The remaining failures are caused by defects in process design, process equipment manufacturing, or plant construction (15.5%), and by external events or natural causes (3.5%). A survey conducted by Mashiguchi of chemical plant failures in Japan has shown a similar distribution [13]. Venkatasubramanium [14] cites studies which state that human error may account for up to as much as 70% of industrial accidents. Although highly anecdotal in nature, such failure distributions do provide a good indication of the classes of failures that are not being addressed properly by current fault management measures. Nimmo cites studies which estimate that the results of abnormal process operations (including inadequate process fault management) cost the U.S. petrochemical industry alone $20 billion per year [15].

The first, and most effective way to eliminate potential process faults is to keep them out of the process system design. As Lees [16] states: "The safety of the plant is determined primarily by the quality of the basic design rather than by the addition of special safety features." Correspondingly, to identify potential operating safety problems, hazardous operation (HAZOP) studies [17] are now commonly performed during the design and construction of both new process systems and process retrofits. Such studies systematically examine alternative designs for potential safety problems, thereby allowing these designs to be compared on a common basis of potential operating risk. They are also very useful tools for determining how that risk will be affected by a particular process system improvement or operating failure. Software tools that perform online risk analysis are also now becoming available [18–31]. Such tools should allow operators to keep apprised of the relative levels of risk associated with operating a process in partially failed modes.

Nonetheless, it will never be possible to totally eliminate the risk inherent in chemical plant operations. The best that can be done is to reduce this risk below an acceptable limit. Moreover, risk analysis studies can be performed improperly, either by overlooking some potential process faults or by using poor estimates of the risk factors associated with particular faults [32–34].

Constructing process plants with higher-quality equipment and employing comprehensive preventive maintenance programs are two methods designed to improve the reliability of process equipment during production runs. Both methods reduce the likelihood that a particular system component will malfunction, thus forcing an emergency process shutdown or causing an accident. Additionally, preventive maintenance sometimes uncovers incipient problems before they develop into major equipment failures that cause long process downtimes. This allows appropriate corrective actions to be taken well before such situations occur. Determining optimal scheduling of preventive maintenance shutdowns is a problem currently under active research [35–38].

Regardless, because of the stochastic nature of equipment failure, accurate prediction of when a particular process system component will malfunction is impossible. Preventive maintenance can therefore not be used to eliminate all process equipment failures. A very good example of this can be seen in the production of ammonia [39, 40]. An average ammonia plant is out of service for preventive maintenance approximately 20 days a year. These plants are still out of service for a similar period of time due to unpredicted or sporadic equipment failures.

The final method of reducing the occurrence of process faults is the establishment of standard operating procedures. Establishing and then following such procedures strictly represents the most straightforward way to reduce process faults caused by human errors. This is because such procedures provide operators with simple guidelines for proper operation of a process system. Distilled from past operating experience and from safety considerations of which the operators may not even be aware, such procedures specify the sequence of actions that have been proven to control process operation effectively and safely. By following such predetermined procedures, the various operators' responses to particular situations will be more predictable, more consistent, and consequently, more reliable.

However, there are several potential problems that can result from relying too heavily on standard operating procedures for safe process operation. These arise from the requirement that the operator remember all of the procedures, together with their exceptions, and then use them in the appropriate situations. For a given situation, operators may not apply the proper procedures because they either have never fully learned or have forgotten the correct procedures. Another problem is that they may misinterpret the situation and apply incorrect procedures. A third potential problem is that operators may be confronted with a situation for which either predetermined procedures do not exist or for which those that do exist are not appropriate. Another potential problem is that operators may ignore predetermined procedures and attempt to devise their own, thereby defeating the purpose of having standard operating procedures. The final potential problem is that they may blindly follow a predetermined procedure even though it leads directly to the development of an emergency situation.

Comprehensively training operators in effective process fault management is the best way to overcome the problems noted above. Training is designed to develop three critical cognitive abilities in operators [41]. The first is to give them knowledge of what the system will do by itself to recover from abnormal operating conditions and what operators are required to do; the second is to give operators knowledge of how a system will respond to unwarranted inputs; and the third is to reduce operators' stress when abnormal process operating conditions occur. As a result of such fault management training, operators develop competence as well as confidence in their ability to perform successfully during abnormal process operating conditions [41].

Nonetheless, problems can arise when attempting to train operators properly in process fault management. Perhaps the greatest problem is that typically it is not possible for operators to practice with, and be tested on, the actual process system during their training. This causes various problems in trying to simulate the actual process conditions [42]. Operators are still expected to be versatile enough to diagnose faults they have not experienced previously, might not fully understand, and perhaps might not foresee [42]. As noted by both Ducan [42] and Rasmussen [43], the operators who are most proficient at fault management are those who are trained by experienced operators or by the experience of controlling the process system manually. Consequently, the current trend toward fewer process operators and automated process control will cause the operators' training to be less than optimal.

Adding alarm systems to a process control system is another method currently used to help operators perform process fault management. Alarm systems are designed to alert operators to abnormal operating conditions before emergency situations develop. Such warnings usually give operators sufficient information to identify, respond to, and completely rectify the cause of abnormal conditions long before an emergency shutdown occurs. Furthermore, even if such a shutdown does occur, the cause can usually be readily determined from the pattern of alarm messages that are generated before the interlock system is activated.

Unfortunately, the proliferation of alarms in chemical and nuclear plants has led to a situation known as *alarm inflation*. This situation has arisen because of the increased number of alarms used in these plants, the fact that many of these alarms do not directly indicate the cause of the abnormal process conditions, and the fact that an inconsistent philosophy is sometimes used to design alarm systems [7]. Nonoptimal alarm design [44, 45] (1) frustrates operators, (2) triggers needless process deviation investigations, and (3) causes others to perceive that the process is not in control. This leads to a fundamental definition of an ideal alarm: that an ideal alarm represents an abnormal condition that requires a response.

Currently, alarm inflation greatly complicates an analysis required by an operator to understand the underlying causes of the alarms. This, in turn, creates two potential problems in process fault management. The first is that operators may ignore significant alarms. In a study of process alarm systems, Kortlandt [5,46] discovered that only 10% of the alarms in one plant caused operators to take corrective action. Of the other alarms, 50% were followed by no operator action whatsoever, while 40% resulted directly as a consequence of a previous operator action. Obviously, in such situations the effectiveness of the alarm systems to alert the operators to process problems is very low. Worse yet, this can lead to a situation in which the operators ignore activated alarms because they believe those alarms to be either unreliable or unimportant. Many plant accidents have resulted from just such operator behavior [10].

Alarm inflation (also referred to as *alarm floods*) [47] adds greatly to the problem known as *cognitive overload*. During major process upsets, hundreds or even thousands of alarms can become activated very quickly. This situation makes it very difficult for an operator to identify the most significant alarms and then diagnose the cause or causes of those alarms [46]. Consequently, alarm inflation defeats the original purpose of an alarm system, which is to help an operator assimilate more effectively the large amount of information coming into the control room [46] into an accurate model of the process state. This has led to efforts to create *smart alarms* [44, 45] or *intelligent alerts*.

These can (1) access all pertinent process information within a process control computer, (2) provide insight to operators as to the root cause, (3) display the expected operator response, and (4) link to paging systems.

Another method currently used to help operators perform process fault management is the addition of emergency interlock systems to a process control system. Since the role played by interlock systems in process safety was discussed earlier, only a few comments about their inherent limitations are given here. Interlock systems can be poorly designed, can be overridden by process operators, can fail, and can even be inconsequential in averting major catastrophes. Therefore, the mere existence of emergency interlock systems does not guarantee that process personnel and equipment will be protected completely.

The final method for helping operators perform process fault management is the creation of better control consoles and human–machine interfaces. Such consoles and interfaces are intended to help operators extract the most crucial information from a background of irrelevant information [6]. Better designs, along with more intelligent interpretation of the process data, appear to be the best means for dealing with problems of information overload. However, the problem with developing general interfaces is that the information operators need depends on the particular situation being confronted at that moment. For example, the information required by operators during process startup is different from the information they use to solve a problem during normal process operation. This makes it very difficult to design interfaces that are optimal for all possible situations.

Despite these efforts, inadequate fault management continues to cause major accidents within the chemical process industries. This is evident by the catastrophic accidents at Bhopal, India (1984), Mexico City, Mexico (1984), and São Paulo, Brazil (1984) [10]. It also represents a major problem for the nuclear power industry, as is evident by the accidents at power plants located at Three Mile Island, Pennsylvania (1979), and at Chernobyl in Ukraine (1987). Although these accidents have been widely publicized, the vast majority of plant mishaps have not. As a result, general lessons that could have been learned from these accidents are either never fully presented or are quickly forgotten [10]. In fact, many accidents have been caused by the same mistakes being repeated over and over. The most general lesson that can be learned from past incidents is that almost all plant accidents are preventable if the emergency situations preceding them are properly recognized and acted upon correctly.[1] At a minimum this requires proper recognition of those emergency

---

[1]According to Duncan A. Rowan, a retired DuPont forensic investigator, the majority of the catastrophic accidents he investigated were caused by simple single-fault situations that were misinterpreted and responded to improperly by process operating personnel.

situations. Unfortunately, even this does not guarantee that fault management will be performed correctly.

1.5   LIMITATIONS OF HUMAN OPERATORS IN PERFORMING PROCESS FAULT MANAGEMENT

The preceding discussion has indicated that the various measures taken to improve process fault management do not always guarantee successful results; accidents still occur. One reason for this is that some of these measures are not always properly implemented or adequately maintained. Even if they are, these measures alone do not provide operators with sufficient support in all emergency situations. Moreover, it is extremely doubtful that measures guaranteed to provide such support can ever be developed. Human beings have certain inherent limitations that always cause their performance as process operators to be potentially unreliable.

One of these limitations is *vigilance decrement.* Studies have shown that human beings do not perform monitoring tasks very well. The number of things that go unnoticed increases the longer a person performs a given monitoring task [48]. With process operators, this phenomenon results directly from fatigue and boredom associated with control room duties in modern production plants. Automation has left process operators with fewer control functions to perform. This leads to both greater job de-skilling and dissatisfaction. This, in turn, causes boredom that leads to inattention. Since an inattentive operator will probably not have an accurate, up-to-date cognitive model of the process state when confronted with an emergency situation, he or she may mistakenly base decisions on an inaccurate model. Studies have also shown that the quality of a decision depends on the amount of time the decision maker has available. In an emergency situation, an inattentive operator will usually be forced to gather data and make decisions in less time than if he or she had been paying attention. Both of these situations will increase the likelihood of human error. Counteracting this limitation requires a means of relentlessly monitoring and analyzing a process state correctly. Since an agent performing such monitoring and analysis would always be aware of the actual process state, the agent would maximize the time available to the decision maker when process operating problems arise.

Another limitation of human operators is the phenomenon of *mind-set* [10], also known as *cognitive lockup, cognitive narrowing* [49], *tunnel vision* [46], and *point of no return* [43]. Sometimes when an operator becomes sufficiently certain of the cause of abnormal process behavior, she or he becomes committed exclusively to that hypothesis and acts upon it accordingly. This commitment continues regardless of any additional evidence the operator receives which refutes that hypothesis or makes alternative hypotheses more plausible. In most cases, this additional evidence is ignored by the operator until it is too late to initiate proper corrective action [49]. Moreover, the longer an operator observes that the response of the system is not as would be expected, the harder that she or he tries to force it to be so [49]. Counteracting this limitation requires a means of examining all the available evidence in a rational, unbiased manner so that all plausible fault hypotheses consistent with that evidence can be derived. These hypotheses would have to be ranked according to how well they explained the process behavior observed, and this ranking would have to be updated as new evidence became available.

A third human limitation is the phenomenon of *cognitive overload.* Even when the detection of system failures is automatic, the sheer number of alarms in the first few minutes of a major process failure can bewilder operators [49]. Rapid transition of the process state may also do this, especially if operators have not experienced a similar situation and have not been told what to expect [4]. Both situations greatly increase the levels of stress experienced by those operators [50]. Under stressful situations, human beings lose information-processing capability. A direct consequence of this loss is that the operator may not be able to analyze the process state quickly and formulate the appropriate corrective response [41]. Counteracting this limitation requires a means for rapid, rational, and consistent analysis of the process state, regardless of how abnormal it is or how quickly it is changing. Such an analysis would focus the operator's attention on the most likely causes of the process behavior observed, rather than having to attempt to imagine all the possible causes of such behavior.

A fourth limitation of human operators is that the situation confronting them may require knowledge that is beyond their ability to understand [9] or outside the knowledge they have gained from their experience and training [10], or knowledge that they have forgotten [10]. Although operators are generally competent, they typically do not fully understand the underlying fundamental principles involved in the process system's design and operation [10]. Such knowledge is required so that the operators are more capable of flexible and analytical thought during emergency situations. This creates the somewhat paradoxical situation of the need for highly trained personnel to operate "automated" plants [6]. Counteracting this limitation requires a medium in which all pertinent information about the process system can be stored permanently and retrieved quickly. It also requires a method of determining which information is relevant to the solution of the problem currently confronting the process operator.

The final human limitation is that, even in the best of situations, humans make errors. Despite efforts intended to reduce such errors, human errors can never be totally eliminated. Sheridan [49] eloquently states the reason why:

> Human errors are woven into the fabric of human behavior, in that, while not intending to make any errors, people make implicit and explicit decisions, based upon what they have been taught and what they have experienced, which then determines error tendencies.

He adds [49]:

> The results of the human error may be subsequent machine errors, or it may embarrass, fluster, frighten, or confuse the person so that he is more likely to make additional errors himself.

Counteracting this limitation requires a means of storing the correct solutions to operating problems confronted in the past, correctly classifying the current plant situation as one of those problems, and then instantiating the appropriate stored solution with the current process state information. This would enable all the proper analyses performed in the past to be reused efficiently and systematically, thereby eliminating the need to recreate them each time they are required. It should also decrease the chances that the wrong analysis would be used or that the correct analysis would be used improperly.

1.6 THE ROLE OF AUTOMATED PROCESS FAULT ANALYSIS

Measures taken in the past to help operators perform process fault management have not been able to provide them with the support that they need for total elimination of process accidents. Typically, such accidents have had very simple origins [10,51]. These accidents have occurred because the number of possible process failures that need to be considered and the amount of process information that has to be analyzed commonly exceed those that an operator can cope with effectively in emergency situations.

Furthermore, this situation probably cannot be counteracted by additional investments in the various measures discussed previously. Many of these methods have already been developed to nearly their full potential. Thus, to further improve process safety, additional process fault management methods need to be developed to help address this problem directly.

An attractive approach to helping operators perform process fault management is to automate the analysis required to determine the cause or causes of abnormal process behavior: that is, to automate process fault analysis.

Not surprisingly, strategies for automating fault diagnosis in chemical and nuclear process plants have been proposed for nearly as long as computers have been used in process control. However, at the present time, the potential of the process control computers to analyze process information for such purposes is still relatively unexploited [14]. The reasons for this are discussed in Appendix A.

Automated process fault analysis should be used to augment, not replace, human capabilities in process fault management. Consider the relative strengths and weaknesses of automated analysis compared with human analysis. Computers can outperform humans in doing numerous, precise, and rapid calculations and in making associative and inferential judgments [49]. On the other hand, people are better at functions that cannot be standardized. They are also better at decision making that has not been adequately formalized (i.e., creative thought) and in coordinations that involve the integration of a great many factors whose subtleties or nonquantifiable attributes defy computer implementation [1]. These differences need to be kept in mind when designing automated fault analyzers.

Currently, the computer offers a means to analyze process information rapidly in a systematic and predetermined manner. If such analysis is already being done by the operators, automating the analysis would free them to perform other functions. If it is not being done, it could be because the operators do not have either sufficient time or the capabilities required. In either case, proper automation of such analysis should make the information reaching the operators more meaningful [2]. Thus, the main advantage of real-time online fault analysis is to reduce the cognitive load on operators [52], to allow them to concentrate on those analyses that require human judgment.

1.7 ANTICIPATED FUTURE CPI TRENDS

The preceding discussion has indicated how automated process fault analysis can be used to help operators better perform process fault management. The main reasons that such automation will continue to increase have to do with the trends predicted for the CPI [53].

Two of these trends are toward an increased emphasis on quality control and process optimization. Both trends will require that a given process system be operated under tighter control limits. It will thus become more important to detect and correct incipient failures long before they cause major operating problems. The sensitivity analysis required to do this will probably have to be much greater than that which can be performed continuously by even the best human operators.

Another trend predicted within the chemical industry is toward more flexible plants operated over a wider range of operating conditions and producing a wider variety of products. This will make it more difficult for operators to perform fault management because they will have had much less operating experience than in the past with any particular process system or set of operating conditions. With less experience on which to base their decisions, it will be more difficult for operators to determine if observed process behavior is normal and to determine the underlying cause or causes when it is not.

Two other trends predicted are toward specialty product manufacturing and shorter product life cycles. Again, both trends will probably make it more difficult for the operators to perform effective fault management because they will have had much less operational experience with these processes.

It is possible that some of the current problems in process fault management could be counteracted by replacing operators with more highly trained process engineers. The higher wages paid to the engineers would be offset by their more effective operation of the process system. Nonetheless, since engineers are subject to the same human limitations as process operators, doing this would only postpone the need to directly address the problems discussed above. Consequently, it is our contention that automating process fault analysis represents the best currently unexploited means available to address these problems directly. A generalized model-based methodology for optimal automation of process fault analysis is the *method of minimal evidence* (MOME), described in detail later. First, we define some of the concepts associated with process fault analysis that are used throughout our discussion.

1.8 PROCESS FAULT ANALYSIS CONCEPT TERMINOLOGY

The terms *fault* and *fault situation* refer to the actual event that is creating either a potential or an actual process operating problem. In contrast, the term *symptom* refers to observable manifestations of fault situations and to observable manifestations of nonfault events. For example, a stuck valve is considered a fault situation whether or not it affects process operation in any observable way. Furthermore, if the valve causes a high-temperature alarm, for example, the alarm is considered a symptom of that fault situation. It might also be a symptom of many other fault situations and nonfault events.

Diagnostic evidence refers to any symptom that supports the plausibility of one or more fault hypotheses. This evidence may also support the plausibility of any number of nonfault hypotheses. Diagnostic evidence is based on information obtained from the current operating behavior of the target process system. This current process information is refined into diagnostic evidence through the application of knowledge regarding a process system's normal operating behavior. It will be assumed that only the values of the process variables measured directly by process sensors at a constant sampling frequency are used to create diagnostic evidence. This assumption is made for convenience and will not affect the generality of the following discussion whatsoever.

A *target process system* is that portion of the entire system being actively analyzed for process fault situations by the fault analyzer.

A *process operating event* is any occurrence that affects a process system's normal operating behavior sufficiently to generate symptoms. Such occurrences can be actual process fault situations such as pump failures or stuck valves, or nonfault events such as normal process startups and shutdowns or production rate changes. Any combination of process fault situations and nonfault events is also considered a process operating event.

A fault analyzer's *intended scope* is that subset of potential process operating events that the fault analyzer is designed explicitly to correctly identify and classify. By definition, a fault analyzer's intended scope includes its *target fault situations*, those potential process fault situations that the fault analyzer is specifically designed to detect and diagnose. The intended scope is constrained by the fault analyzer's *intended operational domain*, which specifies the process operating states that the fault analyzer can analyze for its target fault situations: for example, all possible process production levels and steady- and unsteady-state operation.

Classifying a process operating event correctly means that the fault analyzer can properly discriminate that event from all other possible events (i.e., the fault analyzer will not misdiagnose that event as another). According to this definition, to operate properly the fault analyzer does not necessarily have to identify a process operating event explicitly; it just has not to misidentify that event. Thus, to classify an event correctly, the fault analyzer has to either diagnose that event as the fault that is occurring or to misdiagnose it by not making any diagnosis whatsoever (i.e., by remaining silent). Performing in this conservative manner is advantageous because it will not distract or confuse operators during emergency situations.

A fault analyzer's competence refers to how correctly it classifies the various possible process operating events contained within its intended scope. A fault analyzer that can classify all of these events correctly is considered *completely competent*.

A fault analyzer's competence can be quite different from its *robustness*, which refers to how correctly a fault analyzer can classify the various process operating events that actually occur during a target process system's operation. Consequently, a fault analyzer's competence is related to how well its performance lives up to its design specifications, whereas its robustness is an indication of how useful those design specifications actually are for the target process system. A fault analyzer's competence will approach its robustness, as its intended scope includes more of the operating events that can actually occur during the process system's operation.

A fault analyzer's *diagnostic resolution* is the level of discrimination between a particular fault situation and all other possible process operating events. *Perfect resolution* refers to those fault situations that can be uniquely discriminated from all other possible process operating events contained within the fault analyzer's intended scope.

A fault analyzer's *diagnostic sensitivity* for a particular fault situation refers to the minimum values of that fault situation's magnitude or rate of occurrence that can be detected and diagnosed correctly. These values depend directly on the current process operating conditions and the degree of diagnostic resolution sought. Fault situations that have magnitudes and/or rates of occurrence at or above a fault analyzer's minimum diagnostic sensitivity are considered *significant* with respect to the fault analyzer.

A fault analyzer's *diagnostic response time* is the interval between the onset of a significant process operating event and its diagnosis by the fault analyzer. It is thus the interval required to expose the complete pattern of symptoms used by the fault analyzer to diagnose its associated faults.

Finally, a fault analyzer's *utility* is a measure of the usefulness derived from its diagnoses by process operators. Utility is thus a composite metric of the fault analyzer's competence and robustness and also its possible diagnostic resolution, sensitivity, and response time for identifying each of its covered process operating events.

REFERENCES

1. Lefkowitz, I., "Hierarchical Control in Large Scale Industrial Systems," in *Studies in Management Science and Systems*, Vol. 7, North-Holland, New York, 1982, pp. 65–98.

2. De Heer, L. E., "Plant Scale Process Monitoring and Control Systems: Eighteen Years and Counting," in *Proceedings of the First International Conference on Foundations of Computer Aided Process Operations*, ed. by G. V. Reklaitis and H. D. Spriggs, Elsevier Science, New York, 1987, pp. 33–66.

3. Syrbe, M., "Automatic Error Detection and Error Recording of a Distributed Fault-Tolerant Process Computer System," in *Human Detection and Diagnosis of System Failures*, ed. by J. Rasmussen and W. B. Rouse, Plenum Press, New York, 1981, pp. 475–486.

4. Linhou, D. A., "Aiding Process Plant Operators in Fault Finding and Corrective Action," in *Human Detection and Diagnosis of System Failures*, ed. by J. Rasmussen and W. B. Rouse, Plenum Press, New York, 1981, pp. 501–522.

5. Rijnsdorp, J. E., "The Man–Machine Interface," *Chemistry and Industry*, May 1986, pp. 304–309.

6. Visick, D., "Human Operators and Their Role in an Automated Plant," *Chemistry and Industry*, May 1986, pp. 199–203.

7. Kohan, D., "The Design of Interlocks and Alarms," *Chemical Engineering*, February, 1984, pp. 73–80.

8. Lees, F. P., "Computer Support for Diagnostic Tasks in the Process Industries," in *Human Detection and Diagnosis of System Failures*, ed. by J. Rasmussen and W. Rouse, Plenum Press, New York, 1981, pp. 369–388.

9. Goff, K. W., "Artificial Intelligence in Process Control," *Mechanical Engineering*, October 1985, pp. 53–57.

10. Kletz, T. A., *What Went Wrong?: Case Histories of Process Plant Disasters*, Gulf Publishing, Houston, TX, 1985.

11. Barclay, D. A., "Protecting Process Safety Interlocks," *Chemical Engineering Progress*, Vol. 84, No. 2, February 1988, pp. 20–24.

12. "Chemical Industry to Act on Safety," *Processing*, December 1985, p. 19.

13. O'Shima, E., "Computer Aided Plant Operation," *Computers and Chemical Engineering*, Vol. 7, No. 4, 1983, pp. 311–329.

14. Venkatasubramanian, V., "Process Fault Detection and Diagnosis: Past, Present, and Future," *Proceedings of CHEMFAS4*, Seoul, Korea, 2001, pp. 3–15.

15. Nimmo, I., "Adequately Address Abnormal Operations," *Chemical Engineering Progress*, September 1995, pp. 36–45.

16. Lees, F. P., *Loss Prevention in the Process Industry*, Butterworth, London, 1980.

17. Duxbury, H. A., and M. L. Preston, "The Process Systems Contribution to Process Safety," in *Proceedings of the First International Conference on Foundations of Computer Aided Process Operations*, ed. by G. V. Reklaitis and H. D. Spriggs, Elsevier Science, New York, 1987, pp. 177–198.

18. Arendt, J. S., and M. L. Casada, "Prisim: An Expert System for Process Risk Management," Paper 82e, presented at the *AIChE Spring National Meeting*, Houston, TX, April 1987.

19. Arueti, S., W. C. Gekler, M. Kazarians, and S. Kaplan, "Integrated Risk Assessment Program for Risk Management," Paper 83e, presented at the *AIChE Spring National Meeting*, Houston, TX, April 1987.

20. Venkatsubramanian, V., and R. Vaidhyanathan, "A Knowledge-Based Framework for Automating HAZOP Analysis," *AIChE Journal*, Vol. 40, No. 3, 1994, pp. 496–505.

21. Vaidhyanathan, R., and V. Venkatsubramanian, "Experience with an Expert System for Automated HAZOP Analysis," *Computers and Chemical Engineering*, Vol. 20, Suppl. B, 1996, pp. S1589-S1594.

22. Suh, J. C., S. Lee, and E. S. Yoon, "New Strategy for Automated Hazard Analysis of Chemical Plants: Part 2. Reasoning Algorithm and Case Study," *Journal of Loss Prevention in the Process Industries*, Vol. 10, No. 2, 1997, pp. 127–134.

23. Venkatsubramanian, V., J. S. Zhao, and S. Viswanathan, "Intelligent Systems for HAZOP Analysis of Complex Process Plants," *Computers and Chemical Engineering*, Vol. 24, No. 9–10, 2000, pp. 2291–2302.

24. Kang, B., E. S. Yoon, and J. C. Suh, "Application of Automated Hazard Analysis by New Multiple Process-Representation Models to Chemical Plants," *Industrial and Engineering Chemistry Research*, Vol. 40, No. 8, 2001, pp. 1891–1902.

25. Zhao, C., M. Brushan, and V. Venkatsubramanian, "Phausuite: An Automated HAZOP Analysis Tool for Chemical Processes: Part I. Knowledge Engineering Framework," *Process Safety and Environmental Protection*, Vol. 83, No. B6, 2005, pp. 509–532.

26. Zhao, C., M. Brushan, and V. Venkatsubramanian, "Phausuite: An Automated HAZOP Analysis Tool for Chemical Processes: Part II. Implementation and Case Study," *Process Safety and Environmental Protection*, Vol. 83, No. B6, 2005, pp. 533–548.

27. Cui, L., J. Zhao, T. Qui, et al., "Layered Diagraph Model for HAZOP Analysis of Chemical Processes," *Process Safety Progress*, Vol. 27, No. 4, 2008, pp. 293–305.

28. Wang, H., B. Chen, X. He, et al., "A Sign Digraphs Based Method for Detecting Inherently Unsafe Factors of Chemical Processes at Conceptual Design Stage," *Chinese Journal of Chemical Engineering*, Vol. 16, No. 1, 2008, pp. 52–56.

29. Zhao, J., L. Cui, L. Zhao, et al., "Learning HAZOP Expert System by Case-Based Reasoning and Ontology," *Computers and Chemical Engineering*, Vol. 33, No. 1, 2009, pp. 371–378.

30. Rahman, S., F. Khan, and B. Veitch, "ExpHAZOP(+): Knowledge-Based Expert System to Conduct Automated HAZOP," *Journal of Loss Prevention in the Process Industries*, Vol. 22, No. 4, 2009, pp. 373–380.

31. Wang, H., B. Chen, X. He, et al., "SDG-Based HAZOP Analysis of Operating Mistakes for PVC Process," *Process Safety and Environmental Protection*, Vol. 87, No. 1, 2009, pp. 40–46.

32. Groen, F. J., C. Smidts, and A. Mosleh, "QRAS: Quantative Risk Assessment System," *Reliability Engineering and System Safety*, Vol. 91, No. 3, 2006, pp. 292–304.

33. Meel, A., and W. D. Seider, "Plant Specific Dynamic Failure Assessment Using Baysian Theory," *Chemical Engineering Science*, Vol. 61, No. 21, 2006, pp. 7036–7056.

34. Meel, A., and W. D. Seider, "Real Time Risk Analysis of Safety Systems," *Computers and Chemical Engineering*, Vol. 32, No. 4–5, 2008, pp. 827–840.

35. Wilson, R. L., Jr., "Back to Basics: Redefining the Mission Planning in an Automated World," in *Proceedings of the First International Conference on Foundations of Computer Aided Process Operations*, ed. by G. V. Reklaitis and H. D. Spriggs, Elsevier Science, New York, 1987, pp. 253–276.

36. Grievink, J., K. Smit, R. Dekker, and C. F. H. van Rijn, "Managing Reliability and Maintenance in the Process Industry," in *Proceedings of the Second International Conference on Foundations of Computer Aided Process Operations*, ed. by

D. W. T. Rippin, J. C. Hale, and J. F. Davis, CACHE, Inc., Austin, TX, 1994, pp. 133–157.

37. Yang, S. K., "A Condition-Based Failure Prediction and Processing Scheme for Preventive Maintenance," *IEEE Transactions on Reliability*, Vol. 52, No. 3, 2003, pp. 373–383.

38. Peng, Y., M. Dong, and M. J. Zuo, "Current Status of Machine Prognostics in Condition-Based Maintenance: A Review," *International Journal of Advanced Manufacturing Technology*, Vol. 50, No. 1–4, 2010, pp. 297–313.

39. Mundo, K. J., "Reliability of Chemical Plants," *Large Scale Systems*, ed. by Y. Haines, North-Holland, New York, 1982, pp. 55–67.

40. Williams, G. P., W. W. Hoehing, and R. G. Byington, "Causes of Ammonia Plant Shutdowns: Survey V," *Plant/Operations Progress*, April 1988, pp. 99–110.

41. Dellner, W. J., "The User's Role in Automated Fault Detection and System Recovery," in *Human Detection and Diagnosis of System Failures*, ed. by J. Rasmussen and W. B. Rouse, Plenum Press, New York, 1981, pp. 487–499.

42. Ducan, K. D., "Training for Fault Diagnosis in Industrial Process Plants," in *Human Detection and Diagnosis of System Failures*, ed. by J. Rasmussen and W. B. Rouse, Plenum Press, New York, 1981, pp. 553–573.

43. Rasmussen, J., "Models of Mental Strategies in Process Plant Diagnosis," in *Human Detection and Diagnosis of System Failures*, ed. by J. Rasmussen and W. B. Rouse, Plenum Press, New York, 1981, pp. 251–258.

44. Alford, J. S., J. Kindervater, and R. Stankovich, "Alarm Management for Regulated Industries," *Chemical Engineering Progress*, No. 4, April 2005, pp. 25–30.

45. Alford, J. S., "Bioprocess Control: Advances and Challenges," *Computers and Chemical Engineering*, Vol. 30, 2006, pp. 1464–1475.

46. Lees, F. P., "Process Computer Alarm and Disturbance Analysis: Review of the State of the Art," *Computers and Chemical Engineering*, Vol. 7, No. 6, 1983, pp. 669–694.

47. Crowe, R., "Abnormal Situation Management," *Chemical Engineering Progress*, April 2002, p. 77.

48. Eberts, R. E., "Cognitive Skills and Process Control," *Chemical Engineering Progress*, December 1985, pp. 30–34.

49. Sheridan, T. B., "Understanding Human Error and Aiding Human Diagnostic Behaviour in Nuclear Power Plants," in *Human Detection and Diagnosis of System Failures*, ed. by J. Rasmussen and W. B. Rouse, Plenum Press, New York, 1981, pp. 19–35.

50. Fortin, D. A., T. B. Rooney, and H. Bristol, "Of Christmas Trees and Sweaty Palms," in *Proceedings of the Ninth Annual Advanced Control Conference*, West Lafayette, IN, 1983, pp. 49–54.

51. Lieberman, N. P., *Troubleshooting Process Operations*, PennWell Publishing, Tulsa, OK, 1985.

52. Laffey, T. J., P. A. Cox, J. L. Schmidt, S. M. Kao, and J. Y. Read, "Real-Time Knowledge Based Systems," *AI Magazine*, Spring 1988, p. 27.

53. Schlenker, R. P., Keynote Address: "Process Industry Automation: The Challenge of Change," in *Proceedings of the First International Conference on Foundations of Computer Aided Process Operations*, ed. by G. V. Reklaitis and H. D. Spriggs, Elsevier Science, New York, 1987, pp. 1–26.

2

METHOD OF MINIMAL EVIDENCE: MODEL-BASED REASONING

2.1 OVERVIEW

A general method for performing automated process fault analysis in chemical and nuclear processing plants has been actively sought ever since computers were first incorporated into process control systems [1]. The motivation for this search has been the enormous potential such automation would have for improving both process plant safety and productivity. However, for a variety of reasons, past attempts at automating process fault analysis have not proven to be very successful. They were much more expensive to develop in terms of effort, resources, and time than was planned originally. More important, the resulting computer code performed below expectations and proved to be very difficult to maintain as the target process systems evolved over time. More robust diagnostic strategies and computer implementations are thus still being actively sought [2–5].

A model-based diagnostic strategy called the *method of minimal evidence* (MOME) [1,6–9] that overcomes many of the practical limitations encountered in other strategies is described here. It evolved from the experience gained during the development, verification, and implementation of an automated fault analyzer for a commercial adipic acid plant. This strategy has since been converted into a generalized algorithm based on fuzzy logic

Optimal Automated Process Fault Analysis, First Edition.
Richard J. Fickelscherer and Daniel L. Chester.
© 2013 John Wiley & Sons, Inc. Published 2013 by John Wiley & Sons, Inc.

reasoning [10] and completely automated: All the user has to specify are the engineering models describing normal process operation and their associated statistics computed from sufficient amounts of actual normal operating data. This allows the more difficult problem of automated process fault analysis to be reduced to the much simpler problem of process modeling. In this chapter we describe model-based reasoning as employed by MOME and the types of engineering models that it uses to perform *sensor validation and proactive fault analysis* (SV&PFA).

2.2   INTRODUCTION

Model-based reasoning is a highly systematic and powerful means of deriving plausible hypotheses as to the causes of abnormal process behavior. It was used to create the knowledge bases of the original FALCON [1,6], original FALCONEER [7,8], and various FALCONEER^{TM} IV [9] knowledge-based systems (KBSs). In this chapter we describe effective primary and secondary models, together with how they are obtained, how their associated requisite modeling assumptions are identified, and how they are verified.

The first step in developing competent model-based fault analyzers is to derive a set of as many linearly independent models of normal process operation as possible. These should accurately describe the behavior of the target process system during malfunction-free (i.e., normal) operation. These models include the normal operating characteristics of the process system components, the functional relationships between those components, the process control strategy, and the underlying fundamental conservation, thermodynamic, and physiochemical principles. The set of modeling assumptions required to derive these models defines the domain in which they predict normal process behavior.

Available process knowledge, current process instrumentation, and associated periodical sampling rates can cause the resulting models to vary from highly quantitative (e.g., engineering equations) to highly qualitative (e.g., experiential heuristics) descriptions of the normal process behavior. Regardless, these process models of normal process operation constitute the primary knowledge source used to derive all possible plant state diagnostic evidence. The diagnostic evidence generated by evaluating these models with actual process data is further analyzed with the patterns of model behavior expected during various possible fault situations (i.e., the *SV&PFA diagnostic rules*) which can logically discriminate among possible process operating events within the fault analyzer's intended scope. The specific patterns of diagnostic evidence used for this discrimination depend entirely on the specific model-based diagnostic strategy actually employed [2,11].

Model-based fault analyzers are thus computer programs that determine which of their SV&PFA diagnostic rules most closely match process behavior currently being observed. Their understanding of process fault situations is thus determined completely by their underlying models of normal operation and the consequent SV&PFA diagnostic rules for identifying those fault situations. Therefore, since the diagnostic evidence used by the diagnostic rules is determined directly from evaluating the underlying process models, the fault analyzer's performance is circumscribed completely by an understanding of the normal process behavior represented within those models.

The following terms are used throughout our discussion:

- *Sensor variable*: any process variable that is measured directly by a process sensor (e.g., such as a process temperature being measured by a corresponding thermocouple).

- *Parameter*: any process variable that is not measured directly but which can be adequately characterized by either a standard value and/or an extreme value under normal process operating conditions (e.g., such as the density of a particular process stream).

- *Modeling assumption variable*: any sensor variable or parameter required to derive a given process model that describes process behavior during normal process operation.

- *Primary model*: any mathematical equation that is useful for describing normal process behavior.[1] It must contain at least one sensor variable and must not contain any process variables that are not measured directly by process sensors or are not parameters. It may contain any number of parameters and may also be time dependent (e.g., include differential terms, integral terms, time delays, etc.).

2.3 METHOD OF MINIMAL EVIDENCE OVERVIEW

The method of minimal evidence (MOME) [1,6] is a diagnostic strategy based on the evaluation of engineering models describing normal operation of the target process system with sensor data. It has been demonstrated to be competent in both an adipic acid plant (Figure 2.1) formerly owned and operated by DuPont in Victoria, Texas and a persulfate plant (Figure 2.2) owned and operated by FMC in Tonawanda, New York.

---

[1]Primary models thus conform to the class of world models as opposed to object models because they refer to a collection of information that characterizes the domain of interest rather than characterizing the content of the signal to be processed. World models provide a source of knowledge that is typically used to check the plausibility of potential interpretations of the signals being monitored. World models are thus used inferentially [12].

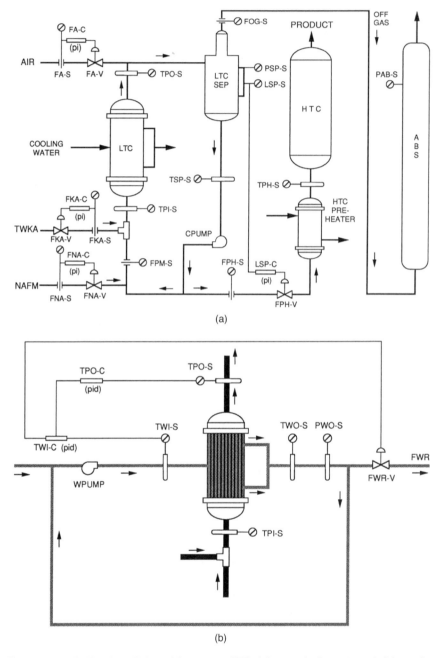

Figure 2.1   DuPont's adipic acid process LTC (a) recycle loop 3 and (b) cooling system.

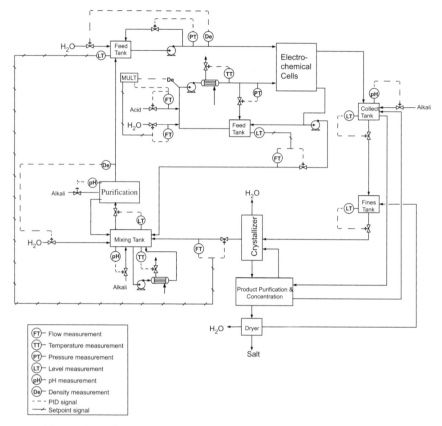

Figure 2.2   Simplified FMC persulfate process with control structure.

The basic logic behind MOME begins with the assumption that once a target process system has been selected for automated fault analysis, as many models as possible that describe the normal operating behavior of that process system should be derived. The models should be based on the most fundamental understanding of normal process behavior available, limited only by the specific type and frequency of the process data being collected. The resulting set of models should constitute a highly accurate description of the target process system's normal operating behavior.

Modeling assumption variables are defined as any process variables about which assumptions must be made to derive a model describing normal operating behavior. Examples of such assumptions include the fact that the value of a sensor measurement used as the value of the variable within the model is correct or that a value of an unmeasured parameter used within the model is equal to its normal and/or extreme value. These assumptions are represented explicitly within the model as assumption variables. Deviations between these assumption variable values and the actual process state generate the residuals (defined below) used by MOME to do sensor validation and proactive fault analysis.

The ith process model describing normal process operation can be characterized as an equation having the following form:

$$0 = f_i \text{ (model } i \text{ modeling assumption variables and time)} \\ - B_i \text{ (process operating conditions)} \tag{2.1}$$

where f_i is a mathematical function and the modeling assumption variables are either specific sensor measurements or standard and/or extreme values of specific parameters, and B_i is the normal offset of model i (referred to as its *beta*) which could be a function of process operating conditions such as current production rate.[2]

Once the modeling assumption variables are instantiated with actual process sensor data or standard and/or extreme values of unmeasured parameters and its current estimated beta has been computed, a residual result (i.e., ε_i below) is just a function of process sensor noise and any currently occurring modeling assumption variable deviations: for example,

$$\varepsilon_i = f_i \text{ (current value of model } i \text{ modeling assumption variables and time)} \\ - B_i \text{ (current process operating conditions)} \tag{2.2}$$

If the residual (i.e., ε_i) of an evaluated model is *significantly high or low* (significantly higher or lower than zero), it can be inferred that at least one or more of the possible modeling assumption variable deviations (i.e., possible process operating events, such as faults) that could cause such a residual is occurring. If the residual (i.e., ε_i) of an evaluated model is not significant, then either (1) there are no modeling assumption variable deviations; (2) one or more such deviations are occurring but at magnitudes or rates of change that are below the sensitivity of that model to discriminate such deviations; or (3) two or more significant assumption variable deviations are interacting in an opposing fashion.

---

[2]Betas can arise for a variety of reasons. Perhaps one or more of the sensor variables used in a residual calculation needs recalibration because of an ongoing measurement bias. This was true for the particular placement of redundant pH meters in one of the FMC FALCONEER™ IV applications. Betas can also arise directly from the particular models used to describe normal operation. In the original FALCON KBS at DuPont, the overall mass balance on the LTC recycle loop had a –7% beta because the HTC preheater feed stream contained 7% offgas, fooling the flow sensor on that stream by that amount. Other causes of betas are from legitimate physical phenomena. In the two FALCONEER™ IV KBSs at FMC, using the Antoine equation to model the pressure/temperature relationship in the crystallizers resulted in a beta equal to the actual boiling-point elevation of the mother liquor.

Assumption variables are classified for each model according to how that model's residual changes with deviations of that particular assumption variable. Modeling assumption variable deviations that cause proportional changes in that given model's residual are defined to be *linear assumption variables*. This is an important distinction because magnitudes of the underlying deviation can be calculated directly from the model's current residual. If the modeling assumption variable is linear in two or more models, these independent estimates of those magnitudes, if they agree, can be used as additional evidence for the diagnosis that the particular assumption variable is indeed deviating; else it is direct evidence against that diagnosis. Alternatively, additional models can be created which directly eliminate those common linear assumption variables. The behavior of these additional linearly dependent (referred to as *secondary*) models' residuals also directly generates such additional diagnostic evidence. The exhaustive use of secondary models' residual behavior is exploited directly by the MOME strategy.

Briefly, the MOME diagnostic methodology for fault analysis compares patterns of primary and secondary residual behavior expected to occur during the various possible assumption variable deviations with the patterns of those residuals currently present in the process. It uses the minimum unique patterns required for correct analysis, allowing for many of the possible multiple assumption variable deviations (i.e., multiple faults) to be identified directly. This is important because this methodology discerns not only process faults but all possible process operating events causing abnormal behavior, fault and nonfault events alike. The logic is designed to venture diagnoses only if it is highly certain of the underlying problem. This conservative behavior is advantageous because it should not confuse its users with implausible diagnoses at times when the actual process operating state is in flux.

The methodology is based on *default reasoning*. If perfect resolution between the various potential assumption variable deviations is possible for the current process operating state, all but one of the various potential assumption variable deviations for a violated model is shown, by other violated and nonviolated models in a diagnostic pattern, not to be deviating. Thus, the only plausible explanation for such a situation is then the remaining assumption variable deviation. If perfect resolution isn't possible for the current process operating conditions, the resulting resolution of the diagnosed faults presented to the operators is the best possible for the current pattern of evidence being generated by the process data. All fault hypotheses presented are logically legitimate explanations of the process behavior observed and need to be considered equally plausible by the process operators.

The patterns of expected residual behavior that result from applying this method (i.e., the SV&PFA diagnostic rules) contain the minimum patterns required to diagnose each of the possible fault situations. This directly maximizes the possible sensitivity of the fault analyzer for these various faults, maximizes the possible resolution (discrimination between various possible faults and nonfault events) of that analysis, and optimizes its overall competence when confronted with multiple assumption variable deviations. The strategy can be used further to determine the strategic placement of process sensors for performing fault analysis and the shrewd division of large process systems for distributing process fault analyzers.

2.3.1   Process Model and Modeling Assumption Variable Classifications

The following describes how both models and assumptions are classified within the framework of the MOME strategy.

2.3.1.1 *Primary Model Classification*   All sets of linearly independent models describing normal operation of the target process system to be monitored are classified as *primary models*. To be useful for sensor validation and proactive fault analysis, these models cannot rely on process variables that are either not measured directly or cannot be adequately characterized by a standard and/or extreme value under normal process operating conditions. The frequency with which the data are being collected must also be sufficient to properly evaluate time-dependent terms in the models (differential, integral, time delays, etc.). It is essential for correct performance by the fault analyzer to determine every modeling assumption on which these models depend and then determine the sensitivity of those models to the various possible modes of the corresponding assumption variable deviations. Furthermore, the variance and normal offset (*beta*) of the residuals must be determined accurately as a function of the current process state (e.g., production rate). If done properly, the resulting primary models are termed *well formulated*. Together this understanding of normal process operations constitutes the *deep knowledge* (i.e., the declarative knowledge) that MOME uses to perform its fault analysis.

Primary models of normal process operation can be derived in a variety of ways. One method is to apply the conservation laws to control volumes surrounding the target process system. Models can also be obtained by deriving semiempirical or empirical correlations between groups of sensor variable measurements during normal process operating conditions. For example, a semiempirical correlation can be derived by relating the sensor variables of a system component via a performance relationship that describes that component's normal operation (e.g., pump curves, control valve correlations, models of controller response). Empirical correlations relating groups of sensor variables can also be obtained by performing data regression. Such correlations can be based on either qualitative physical reasoning or merely consistent observations of process behavior under normal operating conditions. The latter method amounts to obtaining process models by formulating observed experiential knowledge about normal process behavior. Regardless of which method is used, all sets of linearly independent models that take the form of equation (2.1) will be defined as *primary models*. To be consistent throughout, all primary models will be written in the following format:

$$0 = \text{input terms} - \text{output terms} - \text{accumulation terms} - \text{beta} \quad (2.3)$$

2.3.1.2 *Secondary Model Classification*   All possible combinations of two primary models that result from eliminating common linear assumption variables (the formal definitions of linear, continuous nonlinear and discrete nonlinear are given in Section 2.3.1.3) constitute the set of linearly dependent *secondary models*. The resulting possible secondary models will provide useful diagnostic evidence for discriminating between assumption variable deviations k and l if and only if the following inequality holds [1]:

$$\frac{\partial \varepsilon_i / \partial a_k}{\partial \varepsilon_j / \partial a_k} \neq \frac{\partial \varepsilon_i / \partial a_l}{\partial \varepsilon_j / \partial a_l} \quad (2.4)$$

where

ε_i and ε_j = residuals of primary models i and j
a_k and a_l = linear modeling assumption variables k and l

The residuals of the resulting useful secondary models can be calculated directly from the residuals of the primary models via the following formula [13]:

$$\varepsilon_{j+k,i} = \varepsilon_j - \frac{\partial \varepsilon_j / \partial a_i}{\partial \varepsilon_k / \partial a_i} \varepsilon_k \quad (2.5)$$

where

ε_j and ε_k = residuals of primary models j and k
$\varepsilon_{j+k,\,i}$ = residual of secondary model formed by combining primary models j and k, eliminating assumption i
a_i = linear modeling assumption variable i

The variances of the resulting useful secondary models can be calculated directly from the variances of the primary models via the following formula [14] for the sum of variances:

$$v_{J+k,i^2} = v_j^2 + \left(\frac{\partial \varepsilon_j / \partial a_i}{\partial \varepsilon_k / \partial a_i} \right)^2 v_k^2 \tag{2.6}$$

where

> v_j and v_k = variances of primary models j and k
> $v_{j+k,i}$ = variance of secondary model formed by combining primary models j and k, eliminating assumption i
> ε_j and ε_k = residuals of primary models j and k
> a_i = linear modeling assumption variable i

2.3.1.3 *Modeling Assumption Variable Classification*   There are three possible types of modeling assumption variable deviations. Each type is based on how a given model's residual changes as a function of that deviation. They are classified as linear, continuous nonlinear, or discrete nonlinear.

Linear assumption variables are for those a_k values for which ε_i does, but $\partial \varepsilon_i / \partial a_k$ does not, depend on a_k (i.e., the magnitude of the residual is directly proportional to the magnitude of the given assumption variable deviation). Models that rely on such assumption variables are said to have analytical redundancy for those deviations. To be considered deviating significantly (as viewed by the perspective of the fault analyzer), deviations of these assumptions must be large enough to cause at least one of the primary model residuals relying on them to be evaluated significantly high (greater than 0) or low (less than 0). This, in turn, defines the best possible sensitivity of the resulting fault analyzer to those assumption variable deviations.

Linear assumption variables can be eliminated directly by algebraically combining pairs of primary models that rely on them. The resulting secondary models will be useful if and only if the inequality given by inequality (2.4) holds. Typical examples of such assumptions are that flowmeter and thermocouple measurements are correct (e.g., if and only if those variables appear as linear terms in the associated model). FALCONEER™ IV does not actually do the algebra to create secondary models automatically but, instead, uses equation (2.5) to calculate their residuals directly and equation (2.6) to calculate their variances from the two parent primary models' first derivatives and variances, respectively. The motivation for deriving all possible secondary models is that even though they are linearly dependent on their parent primary models, they provide additional and incisive diagnostic evidence when the linear assumption variable eliminated is invalid; the residuals of the two primary models depending on that assumption variable will deviate accordingly while the secondary model should not be affected (i.e., the residual should be close to zero). This is in effect a cross-check that the amounts of deviation observable in the two parent primary models are consistent with the hypothesis that the linear assumption variable eliminated is the cause of those primary model residuals.

Continuous nonlinear assumption variables are for those a_k values for which $\partial \varepsilon_i / \partial a_k$ is continuous and depends on a_k (i.e., the magnitude of the residual is not directly proportional to the magnitude of the given assumption variable deviation). These must also deviate significantly in at least one primary model before the resulting fault analyzer can detect their presence.

Continuous nonlinear modeling assumption variables can be eliminated directly only by the formation of secondary models in a situation where that assumption variable is linear in one of its two parent primary models. FALCONEER™ IV does not currently create these secondary models automatically; they must be configured manually and their statistics evaluated just as their parent primary models are evaluated. They can be eliminated indirectly by the formation of secondary models if they are associated with one or more coupled linear assumption variables common to the original pair of primary models. That is, the various types of assumption variables exist as part of a grouping of terms that are different only by a multiple within the two parent primary models. Typical examples of continuous nonlinear assumption variables are leaks that are assumed to be absent within the tube side of a heat exchanger or the reading of a pH meter that is assumed to be calibrated correctly.

Discrete nonlinear assumption variables are for those a_k values for which $\partial \varepsilon_i / \partial a_k$ is discontinuous and depends on a_k. Any deviations in these assumption variables create immediate disruptions in normal operation of the target process system. Thus, all models that depend on such assumption variables are violated whenever they occur, making all deviations of these assumption variables significant. These assumption variables cannot be eliminated either directly or indirectly by the formation of secondary models; all resulting secondary models depend on them. Examples of such assumption variables are variables that indicate the presence or absence of complete pump failures or interlock activations.

2.3.2 Example of a MOME Primary Model

Example 2.1   This example demonstrates how the modeling assumptions required to derive the models of normal process operation are classified into one of the three possible categories described above. The reason for

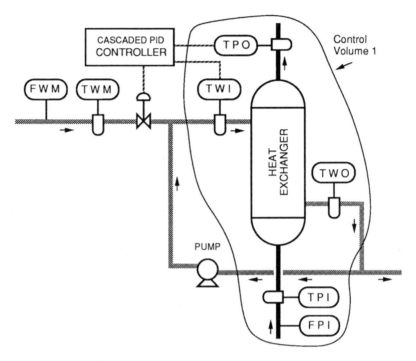

Figure 2.3 Instrumented heat exchanger.

classifying modeling assumption variables as described will become more evident when the SV&PFA diagnostic rule formats are described in Chapters 3 and 4. This example model is based on DuPont's Victoria, Texas adipic acid process system No. 3.[3]

Consider the target process system depicted in Figure 2.3. It consists of a heat exchanger and a cooling-water recycle loop. An energy balance can be derived for control volume 1 if the flow rate of cooling water through the heat exchanger is known. It will be assumed that this flow rate (F_{Wex}) is constant when the water recirculation pump is operating normally. Using

[3]This model was contained in their knowledge-based system (KBS) [a.k.a., the FALCON fault analysis consultant system]. This fault analyzer was tested with over 6000 hours of both actual and simulated process data and was then used online at this plant for approximately three months before the FALCON project was terminated. The FALCON project was a joint venture between DuPont, The Foxboro Company, and the University of Delaware to identify the general issues involved in developing knowledge-based systems for real-time online process fault analysis and then to derive a generalized procedure for doing so. MOME is that generalized procedure. A detailed account of the FALCON project is given in Appendix B.

this modeling assumption, the dynamic energy balance derived for control volume 1 is given by primary model 0 (**P₀**):

$$\mathbf{P_0}: \quad 0 = F_{Pi}\, cp_P(T_{Pi} - T_{Po}) - F_{Wex}cp_W(T_{Wo} - T_{Wi})$$

$$- M_{Wex}cp_W \frac{d(T_{Wo})}{d(\text{time})}$$

The modeling assumption variables required to derive this energy balance include:

1. The sensors for flow rate F_{Pi} and temperatures T_{Pi}, T_{Po}, T_{Wo}, and T_{Wi} are all measuring the process variables accurately.
2. The temperature T_{Wo} accurately represents the bulk temperature of the water within the heat exchanger.
3. The heat capacity coefficients of the process of the process fluid (cp_P) and the cooling water (cp_W) are known and constant (and further that the fluid flowing within the heat exchanger tubes is indeed the normal process fluid and the fluid flowing through the cooling system is indeed water).
4. The amount of water within the heat exchanger (M_{Wex}) is known and constant.
5. The net accumulation of energy within the heat exchanger tubes is insignificant compared to the other terms in the model.
6. No leaks are occurring in either the process or cooling-water piping.
7. The water recirculation pump is operating normally with a known and constant water flow rate (F_{Wex}) through the shell side of the heat exchanger.

All of the assumption variables associated with this model are classified as linear except for a specific type of process leak and one specific mode of recirculating pump failure. The assumption variables concerning the absence of process leaks will be examined first.

In this system, leaks can occur in either the process piping or the cooling-water piping. If the process leak occurs in the tubes of the heat exchanger, it allows cooling water to enter the process stream. For such leaks, the interchange between the two systems would cause the actual process flow through the tube side of the heat exchanger (F_{Pi}) to be greater than that indicated by the upstream flow sensor. Correspondingly, the water flow rate through the shell side of the heat exchanger (F_{Wex}) would drop below its assumed normal value. Also, the contamination of the process fluid would alter its heat capacity coefficient (cp_P). All three of these phenomena would cause the energy balance above to be evaluated as lower than zero.

Although energy would always continue to be conserved in the process system while such tube leaks were occurring (a fundamental conservation law of nature), the form of the model would not be a valid representation of this balance during such faults. In contrast, significant leaks occurring elsewhere in the process system would cause this model to be violated, but its form would remain a valid model of the actual energy balance occurring in the heat exchanger. It is thus not possible to rearrange P_0 to calculate the severity of the tube leaks directly from the energy balance's residual as can be done for leaks occurring elsewhere in the process system. The assumption concerning the absence of process leaks can thus be further classified into two unique assumptions: the absence of those leaks that occur in the tubes of the heat exchanger (a continuous nonlinear assumption) and the absence of those that occur elsewhere in the process system (a linear assumption).

The other assumption variable deviation that would invalidate the form of this model from being a valid representation of the actual energy balance would be the sudden and complete failure of the water recirculation flow. Such a pump failure would violate the assumption that the water flow rate through the shell side of the heat exchanger (F_{Wex}) was equal to its normal, constant value. The locations of the thermocouples are such that the accumulation of energy within the cooling water system cannot be determined accurately by this model if the water circulation ceases completely. Consequently, the model cannot be rearranged for accurate calculation of the flow rate of water through the shell side of the heat exchanger, even though such a pump failure would make it zero. Complete pump failure is classified as a discrete nonlinear assumption. As shown in Figure 2.4 (an actual calculation of the model's residual during such a pump failure in the adipic acid process), the associated models' residuals are instantaneously violated significantly whenever discrete nonlinear assumption variable deviations occur. Such failures could consequently cause the energy balance to be evaluated as either higher or lower than zero.

A complete pump failure needs to be compared to the situation in which the efficiency of the pump degrades slowly over time, thereby gradually reducing the actual water recirculation flow. The latter situation would also invalidate the modeling assumption that the water flowing through the shell side of the heat exchanger was constant and equal to F_{Wex}. However, the model described above could be rearranged for a direct accurate estimation of the reduced water recirculation flow rate (i.e., there is analytical redundancy for this potential fault in this model). Consequently, the assumption concerning the normal operation of the recirculation pump can be classified further into two unique assumptions: the absence of a complete failure of the pump (a discrete

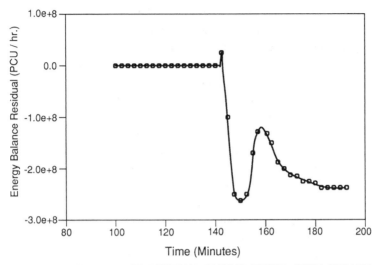

ENERGY BALANCE RESIDUAL DURING WATER PUMP FAILURE

Figure 2.4   DuPont adipic acid LTC cooling water pump failure.

nonlinear assumption) and the absence of a gradual failure in the efficiency
of the pump (a linear assumption).

Primary model $\mathbf{P_0}$ can be modified to make the various assumption variables described above into explicit terms within it.

$\mathbf{P_0^*}$: $\varepsilon_{P0}^* = (F_{Pi} - L_{Pup})c_{PP}T_{Pi} - (F_{Pi} + L_{Pdn})c_{PP}T_{Po}$

$$- \exp(P_{FAIL})\frac{P_{eff}}{100.0}\ln(L_{TUBES})[(F_{Wex} + L_{Wdn})c_{PW}T_{Wo}$$

$$-(F_{Wex} - L_{Wup})c_{PW}T_{Wi}]$$

$$-M_{Wex}c_{PW}\frac{d(T_{Wo})}{d(\text{time})}$$

where

$L_{Pup} = 0$ and assumes no leaks in the process piping upstream of the
heat exchanger (actual leaks are greater than zero)

$L_{Pdn} = 0$ and assumes no leaks in the process piping downstream of the
heat exchanger (actual leaks are greater than zero)

$L_{Wup} = 0$ and assumes no leaks in the cooling water piping upstream of
the heat exchanger (actual leaks are greater than zero)

$L_{Wdn} = 0$ and assumes no leaks in the cooling water piping downstream
of the heat exchanger (actual leaks are greater than zero)

$L_{TUBES} = e$ and assumes that there are no water leaks in the cooling water
tubes (actual tube leaks are less than e)

$P_{eff} = 100.0$ and assumes that cooling water pump efficiency is 100% (actual pump efficiency ranges between 0 and 100%)

$P_{FAIL} = 0.0$ and assumes that the water pump does not fail suddenly and completely (actual such pump failures can cause P_{FAIL} to go either higher or lower than 0.0)

Primary model $\mathbf{P_0}^*$ would be evaluated with the specific current sensor values and normal and/or extreme values of the parameters indicated to determine whether or not all these modeling assumption variables are deviating. It should be noted that the terms modeling the two nonlinear assumptions (i.e., variables L_{TUBES} and P_{FAIL}) are not exactly the actual relationship of how the energy balance will be affected by such assumption variable deviations. This does not represent a problem for our methodology: We do not require accurate fault models for nonlinear assumptions. All that is necessary for our analysis is that these assumption variables are represented as nonlinear terms in the primary model and would always cause the resulting residual to deviate in the proper direction whenever those nonlinear assumption variables deviate significantly.

2.3.3 Example of MOME Secondary Models

Example 2.2 Consider the instrumented mixing tee depicted in Figure 2.5. The following mass and energy balances (primary models $\mathbf{P_1}$ and $\mathbf{P_2}$ below, respectively) can be derived for this system:

Primary Model 1. $\mathbf{P_1}$ is a mass balance on the mixing tee:

$$\mathbf{P_1}: \quad \varepsilon_{P1} = [p1(F1 - L1) + p2(F2 - L2) - p3(F3 + L3)](60\,min/h)$$

where

$$p1 = p2 = p3 = \text{density of water} = 8.34\,lb/gal$$

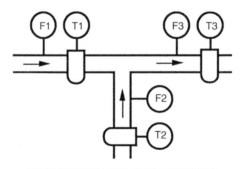

INSTRUMENTED PROCESS MIXING TEE

Figure 2.5 FMC ESP process barometric condenser water feed.

Its derivation requires six modeling assumptions.

Assumption variables a_i for primary model $\mathbf{P_1}$:

a_1:   Linear; F1 (H_2O flow rate into HTX1) flow sensor is correct:
$\partial \varepsilon_{P1}/\partial a_1 = +500$

a_2:   Linear; F2 (H_2O flow rate into HTX2) flow sensor is correct:
$\partial \varepsilon_{P1}/\partial a_2 = +500$

a_3:   Linear; F3 (H_2O flow rate into Cond.) flow sensor is correct:
$\partial \varepsilon_{P1}/\partial a_3 = -500$

a_4:   Linear; L1 = 0; no process leaks in HTX1 water return line:
$\partial \varepsilon_{P1}/\partial a_4 = -500$

a_5:   Linear; L2 = 0; no process leaks in HTX2 water return line:
$\partial \varepsilon_{P1}/\partial a_5 = -500$

a_6:   Linear; L3 = 0; no process leaks in cond. water feed line:
$\partial \varepsilon_{P1}/\partial a_6 = -500$

Primary Model 2. $\mathbf{P_2}$ is an energy balance on the mixing tee:

$$\mathbf{P_2}: \quad \varepsilon_{P2} = [p1 \cdot cp1(F1 - L1)(T1 - 0) + p2 \cdot cp2(F2 - L2)(T2 - 0)$$
$$- p3 \cdot cp3 \cdot (F3 + L3)(T3 - 0)](60\,\text{min/h})$$

where

$p1 = p2 = p3 =$ density of water = 8.34 lb/gal
$cp1 = cp2 = cp3 =$ heat capacity of water = 1.8 Btu/lb $\cdot \,^{\circ}$C

Its derivation requires nine modeling assumptions.

Assumption variables a_i for primary model $\mathbf{P_2}$:

a_1:   Linear; F1 (H_2O flow rate into HTX1) flow sensor is correct:
$\partial \varepsilon_{P2}/\partial a_1 = +900T1$

a_2:   Linear; F2 (H_2O flow rate into HTX2) flow sensor is correct:
$\partial \varepsilon_{P2}/\partial a_2 = +900T2$

a_3:   Linear; F3 (H_2O flow rate into Cond.) flow sensor is correct:
$\partial \varepsilon_{P2}/\partial a_3 = -900T3$

a_4:   Linear; L1 = 0; no process leaks in HTX1 water return line:
$\partial \varepsilon_{P2}/\partial a_4 = -900T1$

a_5:   Linear; L2 = 0; no process leaks in HTX2 water return line:
$\delta \varepsilon_{P2}/\delta a_5 = -900T2$

a_6:   Linear; $L3 = 0$; no process leaks in cond. water feed line:
$\partial \varepsilon_{P2}/\partial a_6 = -900T3$

a_7:   Linear; T1 (H_2O temp. out of HTX1) is correct:
$\partial \varepsilon_{P2}/\partial a_7 = +900(F1 - L1)$

a_8:   Linear; T2 (H_2O temp. out of HTX2) is correct:
$\partial \varepsilon_{P2}/\partial a_8 = +900(F2 - L2)$

a_9:   Linear; T3 (H_2O temp. into TC Cond.) is correct:
$\partial \varepsilon_{P2}/\partial a_9 = -900(F3 + L3)$

From these two primary models, three unique secondary models (S_1, S_2, and S_3 below) can be derived by eliminating flow rates F2, F1, and F3, or leaks L2, L1, and L3, respectively.

Secondary Model S_1. S_1 is a combined mass and energy balance on the mixing tee, eliminating either coupled assumption variable F2 or L2.

S_1: $\varepsilon_{S1} = (F1 - L1)(T1 - T2) - (F3 + L3)(T3 - T2)$

Secondary Model S_2. S_2 is a combined mass and energy balance on the mixing tee, eliminating either coupled assumption variable F1 or L1.

S_2: $\varepsilon_{S2} = (F2 - L2)(T2 - T1) - (F3 + L3)(T3 - T1)$

Secondary Model S_3. S_3 is a combined mass and energy balance on the mixing tee, eliminating either coupled assumption variable F3 or L3.

S_3: $\varepsilon_{S3} = (F1 - L1)(T1 - T3) - (F2 - L2)(T3 - T2)$

The usefulness of the possible secondary models depends on how well inequality (2.4) holds. The limitations placed on the usefulness can be demonstrated as follows.[4]

---

[4]Forming identical secondary models by eliminating two or more unique linear modeling assumption variables from the same parent primary models is referred to as *coupled assumption elimination*. This occurs because the grouping of coupled assumption variables exists in both parent primary models as multiples of identical terms. Although these multiple copies of the same secondary model that result from the coupled assumption elimination do not provide additional evidence to discriminate between the coupled assumption variable deviations, they do provide evidence for other possible assumption variable deviations. Consequently, although redundant, they do not interfere with or impair the subsequent fault analysis based on them and the other primary and secondary model residuals. In FALCONEER™ IV, all possible secondary model residuals are created serially and their results inferred exhaustively.

Consider the assumption variables stating that flow meters F1 and F2 are operating properly. Taking the ratios of the partial derivatives as indicated by the inequality results in the following inequality:

$$\frac{\partial \varepsilon_{P2}/\partial a_1}{\partial \varepsilon_{P1}/\partial a_1} \neq \frac{\partial \varepsilon_{P2}/\partial a_2}{\partial \varepsilon_{P1}/\partial a_2}$$

$$\frac{+900\text{T}1}{+500} \neq \frac{+900\text{T}2}{+500}$$

$$\text{T}1 \neq \text{T}2$$

As process operating temperatures T1 and T2 approach one another, the usefulness of secondary models S_2 and S_1 in providing useful additional evidence diminishes rapidly [i.e., the same diagnostic evidence given by the mass balance's residual (ε_{P1}) would also be given by the energy balance's residual (ε_{P2}), making further discrimination between errors in flow measurements F1 and F2 impossible].

2.3.4 Primary Model Residuals' Normal Distributions

Determining how accurately each primary model describes actual process behavior is important for interpreting their associated residuals properly.[5] This accuracy establishes the diagnostic sensitivity of the resulting fault analyzer for each possible fault situation. The accuracy of a process model is determined by evaluating that model with actual process data. These data need to be collected over the entire range of normal process operating conditions (e.g., various production rates) in which the model is expected to be satisfied. This allows a representative distribution of that model's residual to be derived as a function of the normal process state.

The resulting distribution for a given model can be used to determine an error tolerance band for that constraint. This error tolerance band is defined by a lower bound, T_i^L, and an upper bound, T_i^H. These bounds for primary model P_i are defined by the following equations:

$$T_i^{\ L} = -\kappa_i^{\ L}\sigma i \tag{2.7}$$

$$T_i^{\ H} = \kappa_i^{\ H}\sigma_i \tag{2.8}$$

[5]Determining exactly which models being evaluated are currently considered violated requires a statistical analysis of the value of their residuals. The analysis required to do so is described here and is based on Boolean logic. Boolean logic greatly simplifies the following discussion of the analytical techniques required to develop competent fault analyzers. These techniques are extended to non-Boolean logic in Chapter 4.

σ_i^2 is equal to the normal variance (i.e., v_i) associated with the distribution of primary model P_i's residual evaluated with normal process operating data [i.e., ε_i as defined by equation (2.2)], and is usually a function of process operating conditions (e.g., process production rate). κ_i^L and κ_i^H are equal to the number of standard deviations used to define the lower and upper normal tolerance limits on the primary model P_i's residual, respectively. The numerical values assigned to κ_i^L and κ_i^H are specific to each model and need to be determined by a statistical analysis of each residual's distribution.

Typically, for models that naturally represent equality relationships (e.g., those derived from conservation laws), the distributions of the models' residuals are normally Gaussian. In such cases, the values of κ_i^L and κ_i^H would be equal. Setting the value of these parameters to 3.0 would be sufficient 99% of the time for defining the typical range of a model's residual. Henceforth, determining that the model's evaluated residual lies within 99% of the residual's normal distribution will be the criterion used to define the break between normal values of the residual and those that are considered significantly abnormal. Furthermore, this criterion will also be used for all process models with other types of residual distributions. However, for residual ε_i distributions that have odd moments about 0.0, the values of κ_i^L and κ_i^H in equations (2.7) and (2.8) will not be equal.

Once the error tolerance band defining the normal range of a given model residual is derived, it is possible to statistically determine the diagnostic significance of any residual. If the residual of primary model P_i is within its associated tolerance band, the residual ε_i is defined as normal (i.e., ε_i^S), and the primary model is defined as satisfied (i.e., P_i^S).[6] If the value of the primary model's residual is greater than the upper bound of its specified tolerance band T_i^H, the residual ε_i is defined as abnormally high and the primary model P_i is defined as violated high (i.e., ε_i^H and P_i^H). Similarly, if the value of the primary model's residual is less than the lower bound T_i^L, residual ε_i is defined as abnormally low and the primary model P_i is defined as violated low (i.e., ε_i^L and P_i^L).

In the following discussions it is assumed that all of the primary and secondary models' residuals are evaluated and that the diagnostic significance of these residuals is determined whenever the sensor variables are sampled. Thus, to perform real-time fault analysis continuously it is necessary to be able to evaluate all of the primary and secondary models' residuals, determine their diagnostic significance, and interpret the resulting patterns of violated and satisfied models within each prescribed data sampling period.

---

[6]The superscripts S, L, and H stand for satisfied, low, and high, respectively.

2.3.5 Minimum Assumption Variable Deviations

The tolerance band associated with a primary model (i.e., T_i^L and T_i^H) defines the range in which that model's residual fluctuates during normal process operation. Therefore, the only explanation for a residual being outside these limits is that the actual plant state is deviating from the state assumed to be present by that model. The violation of a model thus provides direct unambiguous evidence that one or more of that model's underlying assumption variables that can cause such behavior are invalid. In contrast, model satisfactions provide only indirect and rather ambiguous diagnostic evidence. The satisfaction of a primary model indicates either that none of the assumption variables have deviated significantly (i.e., significant for that particular model) or that two or more of the assumption variables have deviated but are interacting in such a way as to cause that model to appear to be satisfied.

Assumption variable deviations are thus classified for each model as either significant or nonsignificant, depending on their magnitudes. The actual process state is considered to be deviating significantly from that assumed to be occurring if that deviation reaches a magnitude that violates one or more models; that is, it causes the evaluated residual of at least one particular primary model to lie outside the error tolerance band associated with that model. If the magnitude of deviation does not cause a given residual to exceed its threshold, that deviation is considered to be nonsignificant with respect to that particular model.[7]

To determine the lowest magnitude of an assumption variable deviation that is significant for a given model, it is necessary to determine the sensitivity of that model with respect to that assumption variable. This can be accomplished by performing a sensitivity analysis as follows.

For linear assumption variables, it is possible to determine directly the theoretical minimum positive and negative assumption variable deviations that need to exist before that model will become violated.[8] These minimum deviations are determined by using the following equations, of which there are two possible cases to consider:

Case 1: $\partial P_i / \partial a_j > 0.0$

$$
\begin{aligned}
d_{i,j}^+ &= \frac{T_i^H}{\partial P_i / \partial a_j} \\[2mm]
d_{i,j}^- &= \frac{T_i^L}{\partial P_i / \partial a_j}
\end{aligned}
\tag{2.9}
$$

---

[7] This definition of significance has no bearing on the severity that that model assumption deviation (e.g., particular process faults) may or may not be having on actual process operation.
[8] The model is said to have *analytical redundancy* for that linear assumption variable.

Case 2: $\partial P_i / \partial a_j < 0.0$

$$\bar{d}_{i,j}^+ = \frac{T_i^L}{\partial P_i / \partial a_j}$$

$$\bar{d}_{i,j}^- = \frac{T_i^H}{\partial P_i / \partial a_j}$$

For a given linear assumption variable a_j, $\bar{d}_{i,j}^+$ represents the largest positive assumption variable deviation that can theoretically exist between the actual process state and the state defined by the process model without forcing the evaluated residual of primary model P_i to be violated [i.e., either P_i^L (case 2) or P_i^H (case 1)]. Similarly, $\bar{d}_{i,j}^-$ represents the largest such negative deviation [i.e., either P_i^L (case 1) or P_i^H (case 2)]. Conversely, these values can be thought of as being the smallest deviations in the linear assumption variables associated with primary model P_i that would still be considered significant. Therefore, the values derived by equation (2.9) are defined as the *theoretical minimum deviations* of those assumption variables with respect to primary model P_i. The equation directly specifies the sensitivity of the model for deviations of that assumption variable.

For assumption variables of which their first partial derivative as defined above is continuous but depends on those assumption variables, the theoretical minimum deviations would have to be back-calculated from the original primary models that require those assumption variables. A positive and negative theoretical minimum deviation of these continuous nonlinear assumption variables can be determined as the appropriate values that cause the corresponding primary model residual to just surpass T_i^L and T_i^H.

Finally, discrete nonlinear assumption variable deviations do not have theoretical minimum deviations; any possible deviation is always considered significant for all primary and secondary models depending on those assumption variables.

The diagnostic logic behind model-based process fault analysis is straightforward. Any significant residual of a given process model can be used directly as evidence that one or more of its associated modeling assumption variables that can cause such behavior has become invalid. Consequently, whenever a model becomes violated, all of its associated modeling assumption variables that can cause such a violation are suspected to have become invalid. Plausible hypotheses as to which assumption variable or variables are actually currently invalid can be inferred by logically interpreting the entire pattern of the relevant residuals' current statuses.[9] Consequently, if any of the modeling

---

[9]Precisely which residuals are considered relevant depends on the specific diagnostic strategy employed.

assumptions of a given model are overlooked, or if the statistical parameters described above are calculated incorrectly, the competency of the resulting fault analyzer will be jeopardized. This has important implications for the usefulness of potential process models as primary models.

As discussed above, a modeling assumption variable becomes invalidated by a significant deviation between the actual process state and that assumed by the model. For most assumption variables, these deviations can occur in either the positive or negative direction. Those caused by a particular assumption variable deviation that is both significant and positive will cause those particular assumption variables to be defined as being invalid and high (identified for assumption variable deviation a_i as a_i^H).[10] Similarly, deviations which are both significant and negative will cause those particular assumption variables to be defined as invalid and low (identified for assumption variable deviation a_i as a_i^L).

The theoretical minimum deviation of a given assumption variable with respect to a given primary or secondary process model is normally not the same for all models. To work properly, the diagnostic rules derived with many of the other diagnostic strategies reported require that the magnitude of their associated assumption variable deviations be large enough to violate all models that rely on those assumption variables. Unfortunately, fault analyzers based on such rules will misdiagnose certain fault situations. Such misdiagnoses may occur if an assumption variable deviates by a magnitude that is large enough to violate some of the models that rely on it, but is insufficient to generate the complete patterns of model violations contained in its associated diagnostic rules. Henceforth such deviations will be referred to as *intermediate assumption variable deviations*.[11]

The only consequence on the performance of the fault analyzer for some intermediate assumption variable deviations is that diagnosis of these deviations will be curtailed and the fault analyzer will not make any announcements. However, some intermediate assumption deviations can cause misdiagnoses. These occur whenever the partial pattern of model violations generated by the intermediate assumption variable deviation exactly matches the pattern used for a unique diagnosis of one of the other fault situations. By eliminating the inconsequential evidence used by other model-based diagnostic strategies,

---

[10]For assumption variables representing sensor measurements, high deviations mean that the sensor reading is higher than the actual value of the phenomenon; for parameters just the opposite is true (i.e., the parameter's value used in the model is lower than the actual phenomenon's value). Low deviations of the assumption variables mean just the opposite for both sensors and parameters.

[11]As discussed above, assumption deviations whose magnitudes and/or rates of occurrence are sufficient to violate a given model are referred to as *significant assumption variable deviations* for that model.

and by fully exploiting all available diagnostic information evaluated from the primary models (i.e., the secondary model evaluations), it is possible to derive optimal patterns of diagnostic evidence (i.e., optimal SV&PFA diagnostic rules) for diagnosing the various process fault situations. The analysis described in Chapters 3 and 4 demonstrates how such optimal diagnostic rules are derived.

2.3.6   Primary Model Derivation Issues

The following examples describe some of the real-world issues concerning the derivation of potential primary models for performing fault analysis.

2.3.6.1   Example of Rate of Occurrence Limits on Model-Based Reasoning

Example 2.3   This example demonstrates the limits of diagnostic sensitivity on detecting and diagnosing faults with engineering models alone. Consider the DuPont adipic acid process shown in Figure 2.1a. An overall mass balance for this system can be derived as follows:

$$\textbf{P3:} \qquad 0 = \text{FKA-S} + \text{FNA-S} + \text{FA-S} - \text{FOG-S} - \text{FPH-S}$$

$$-K_{\text{SEP}}\frac{d(\text{LSP-S})}{d(\text{time})}$$

Now consider that there is a slow drift in the LTC separator-level sensor LSP-S. The PI controller on this level would maintain its setpoint, but the actual separator would either fill or drain, depending on the actual direction of the drift. If this drift was below the diagnostic sensitivity of P3 to detect, the fault would go unnoticed until additional evidence of the fault situation manifested itself. In the adipic acid process, the result would be one of two cases.

Case 1. The actual LTC separator level gets too low. The FPM-S and FPH-S meter readings would become highly erratic as more and more offgas becomes entrained in the recycle and product streams, respectively, cavitating the process pump.

Case 2. The actual LTC separator level gets too high. The FOG-S meter reading would become highly erratic as more and more process liquid becomes entrained in the offgas and hits the associated flowmeter.

Example 2.3 demonstrates the need to augment MOME logic with heuristics based on sensor trend analysis. These heuristics are intended to fill in the gaps in the SV&PFA model residual calculation sensitivities resulting from the normal background noise inherent in the various sensor measurements and process. In FALCONEER™ IV, all process variables and performance equations (see Example 3.2) can have exponentially weighted moving averages (EWMAs) calculated continuously to determine whether or not those variables are in control (see Chapter 6). Erratic behavior of FPM-S, FPH-S, and/or FOG-S sensors would be detected easily by performing these EWMA calculations continuously.

2.3.6.2 Example of Data Sampling Limits on Model-Based Reasoning

Example 2.4   This example demonstrates the limits of diagnostic sensitivity on detecting and diagnosing faults with engineering models alone. Again consider the adipic acid process shown in Figure 2.1b. The LTC process side exit temperature (i.e., TPO-S) is controlled by a cascaded PID controller scheme. The controller output on the master loop is calculated using the following equation:

$$TPO_{error} = TPO\text{-}S - TPO\text{-}SP$$

P4: $$0 = TWI\text{-}SP - \left(K_p \cdot TPO_{error} + \int (K_i \cdot TPO_{error}) d(time) \right.$$
$$\left. + K_d \frac{d(TPO_{error})}{d(time)} \right)$$

Simulating the behavior of a PID feedback controller with P4 (representing the most fundamental description of the PID controller's normal behavior) would be possible only if the controller's input variables (controlled temperature TPO-S, setpoint TPO-SP, and controller constants K_p, K_i and K_d) and output variables (setpoint TWI-SP) were all being measured. In addition, the controlled variable would have to be sampled frequently enough so that the integral and derivative action of the controller could be closely approximated. Without an adequate sampling frequency, the agreement between the actual and simulated responses would be poor during highly transient process operating conditions. Accurate monitoring of whether or not the controller was operating properly would thus be difficult.[12]

---

[12]In the original FALCON KBS application, actual process operating data were sampled every 15 seconds. This interval was not frequent enough to accurately calculate the derivative term in the PID controller model, so this primary model was not used by the KBS. However, primary models of all four PI controllers were included in the KBS.

2.3.6.3 Example of Formulating Deep Knowledge from Heuristics

Example 2.5   A heuristic that was used by process operators of the DuPont adipic acid process (Figure 2.1a) was that under normal circumstances the temperature rise between the LTC separator and the LTC reactor (i.e., TSP-S − TPO-S) was normally about $+3°C$. A rigorous steady-state reactor model based on the known kinetics of adipic acid formation and the treatment of the overhead pipe to the separator as a plug flow reactor and the separator itself as a CSTR reactor leads to the following primary model:

P5:

$$1.0 - X = \cfrac{\dfrac{c p_P(\text{TSP-S} - \text{TPO-S})\exp(-K \cdot \text{fact} \cdot k((\text{TSP-S} + \text{TPO-S})/2))}{(\text{FKA-S} + \text{FPM-S})/\text{rho}_P}}{\dfrac{1.0 + k(\text{TSP-S})(\text{LSP-S} \cdot A + B)}{(\text{FKA-S} + \text{FPM-S})/\text{rho}_P}}$$

where

k (temperature) = first-order reaction-rate constant for that temperature

$c p_P$ = process liquid heat capacity

rho_P = process liquid density

fact = fraction of overhead pipe containing process liquid

$$= \frac{(\text{FPM-S} + \text{FKA-S})/\text{rho}_P}{1.0 + \text{FOG-S} \cdot C((\text{TPS-S} + 273)/(\text{PSP-S} + 14.7))}$$

K, A, B, C = process specific constants

X = overall TWKA conversion factor, which can range from 0.0 to 1.0, a value of 1.0 indicates agreement between the various measurements used in the calculation, especially the temperature rise of (TSP-S − TPO-S)

This model checks the actual temperature rise very precisely during all steady-state process operating conditions based on the underlying "deep knowledge" (i.e., in this case, the engineering models of underlying reactor configuration).

The potential usefulness of a given primary model for detecting assumption variable deviations should always be considered carefully during its derivation. For example, although it is possible to derive models codifying experiential heuristics, the usefulness of such models' residuals for producing diagnostic evidence is usually very limited. First, all of the models' underlying modeling assumption variables may not be fully understood. As a general rule, qualitative process models (i.e., those based on experiential heuristics or by creating confluences through envisionment rather than employing first principles) require many more modeling assumptions, many of which are implicit, to specify the process states in which the model correctly describes normal process behavior. This observation has also been made by Koton [15]. Typically, the more implicit modeling assumptions required, the more likely it is that some of those assumptions will be overlooked. Overlooked assumptions lead directly to misdiagnoses by the fault analyzer whenever they deviate significantly. Second, even if all the modeling assumption variables associated with a given qualitative process model are known, determining the mathematical relationship that exists between all these assumption variables is usually more difficult. Finally, the resulting qualitative model residual's normal variance typically is also larger than the corresponding first-principles model used to describe a given phenomenon. This reduces the sensitivity of the resulting fault analyzer for those associated assumption variable deviations.

Consequently, to generalize the lessons of Examples 2.4 and 2.5, the following philosophy should always be followed when formulating primary models of normal process behavior. With the only limits being the precise variables being measured and the frequency at which they can be sampled by FALCONEER^{TM} IV, the most fundamental models of normal operating behavior should always be derived and evaluated by MOME. This brings to bear the best relationships (i.e., covers the most normal operating situations) that hold between the various assumption variables under normal operation. These models can be as complex as needed. FALCONEER^{TM} IV can then, to the highest extent possible, utilize the engineering knowledge used originally to design those processes for intelligent monitoring of their daily operation.

2.3.7 Method for Improving the Diagnostic Sensitivity of the Resulting Fault Analyzer

It is sometimes possible to improve the fault analyzer's diagnostic sensitivity for its various assumption variable deviations by putting a first-order lag filter on the various model residuals used by the program. This mitigates the associated sensor noise in calculation of the filtered residual. First-order lag filters are computed using the following formula:

$$\Delta X = \alpha(X - !X_{\text{filtered}}) \tag{2.10}$$

$$X_{\text{filtered}} = X + \Delta X \tag{2.11}$$

where

$$!X_{\text{filtered}} = \text{previous value of } X_{\text{filtered}}$$

$\alpha = \text{first-order lag filter magnitude } (\alpha = 1.0 \text{ is the linear extrapolation of } X; \alpha = 0.0 \text{ is no filter action})$

$$X_{\text{filtered}} = \text{filtered value of } X$$

The price for this extra sensitivity is the time lag required from the moment a fault occurs in the process until the filter's buffer is flushed with current process data and the necessary diagnostic symptoms are created (i.e, diagnostic response time is traded for improvement of diagnostic sensitivity). Consequently, fast-acting process faults which quickly cause interlock shutdowns may thus be too fast for the fault analyzer to identify with the more sensitive filtered residuals. Such filters were used in the FALCON system KBS to improve diagnostic sensitivity for the associated fault situations. Performing SV&PFA in FALCONEER^{TM} IV with filtered residuals is currently handled by allowing the user to select an appropriate α for each primary model (the default α value is 0.8) and using an α value of 0.8 for all secondary models generated automatically. Better performance has been encountered when the requisite primary Model statistics are calculated from normal operating data without this associated filtering. Filtering by the program in daily use then tends to make the resulting analysis more conservative [and consequently, less erratic (chatty) in the diagnoses reported].

2.3.8 Intermediate Assumption Deviations, Process Noise, and Process Transients

As discussed in Section 2.3.5, if the magnitude of an assumption variable deviation is below the fault analyzer's diagnostic sensitivity for perfect resolution of that assumption variable deviation, only some of diagnostic evidence required to diagnose that deviation uniquely will be generated. Such assumption variable deviations, called intermediate assumption variable deviations, can cause a fault analyzer to misdiagnose if the pattern of diagnostic evidence that it generates completely matches any of a fault analyzer's other SV&PFA diagnostic rules.

These types of misdiagnoses do not occur in the SV&PFA diagnostic rules created by MOME because all possible patterns for each possible fault are developed. Lower resolution is traded for the higher diagnostic sensitivity needed to identify intermediate assumption deviations, as demonstrated in Example 3.1. For the structured SV&PFA diagnostic rules discussed in Examples 3.1 to 3.4, such misdiagnoses will not occur.

Similar types of misdiagnoses can also be caused by fluctuations in the various constraints' residuals due to normal process measurement noise. Because of these fluctuations, even with filtering it is possible that a model residual can be judged on one inference cycle to be violated and on the next to be satisfied, and vice versa. Such fluctuations will thus alter the pattern of diagnostic evidence being generated by a particular fault. Non-Boolean reasoning methods for counteracting such process measurement noise are discussed in greater detail in Chapter 4.

Another reason that only a portion of the diagnostic evidence may be generated at a particular time during an actual fault situation is because of the various unit operation time constants and transport time delays inherent in the process system. If the time required for a given fault situation to affect the various process models is not the same for all those models, the pattern of diagnostic evidence generated will be a function of time. Misdiagnoses similar to those described above may occur with other diagnostic strategies, while the full pattern of diagnostic evidence develops over time. However, SV&PFA diagnostic rules based on MOME again trade diagnostic resolution for diagnostic response time to eliminate all such misdiagnoses.

2.4 VERIFYING THE VALIDITY AND ACCURACY OF THE VARIOUS PRIMARY MODELS

Once the set of all the primary models has been derived, it is necessary to verify that all the models represent accurate descriptions of the normal process operating behavior. This does not mean that the agreement between the models' predictions and the actual process operating behavior has to be exact. It is only necessary to thoroughly understand how any existing disagreement behaves as a function of normal process operation, such as a function of the production rate. Such information is required to ensure that the statistical analysis described in Section 2.3.4 is performed correctly.

Representing the *declarative knowledge* (also commonly referred to as *domain knowledge*) about the target process system's normal operating behavior as a set of primary models has a major advantage when it comes to verifying that knowledge's validity and accuracy. This is because each of the primary models can be verified independent of all the others. Thus, a team of process engineers familiar with the target process system can scrutinize each model independently to ensure (1) that all of the modeling assumptions required to derive that model are taken into account, (2) that the standard or extreme values of all parameters contained within that model are reasonable, and (3) that the model accurately predicts normal process behavior. If any discrepancies are discovered, the model can be modified (1) by changing its form through the addition or deletion of terms; (2) by adjusting the standard values of its associated parameters; (3) by constraining its domain of application through the modification, addition, or deletion of modeling assumption variables; or (4) by adjusting its tolerance limits. The primary models resulting from this phase of the verification procedure should always predict the actual process behavior accurately during normal process operation.

It is also necessary to verify that the appropriate models become violated whenever one or more of the associated assumption variables becomes significantly invalid. This ensures that the effects of these assumption variable deviations on the behavior of the various models' residuals has been determined correctly.

Verifying that each of the models becomes violated by a significant deviation in any of its associated assumption variables can be done in a variety of ways. The best way is to evaluate the process models with process data collected during actual process operating events, especially fault situations. When evaluated with these data, models relying on any invalidated assumptions should all be violated appropriately. However, the availability of such data is usually limited to only a small fraction of the possible process operating events contained within the fault analyzer's intended scope. This makes it necessary to use other methods to generate the desired process fault data.[13]

---

[13]As discussed in Appendix B, the availability of actual process fault data for the original FALCON project's target adipic acid process system was very limited. Its scarcity motivated the creation of the dynamic simulation model. Fortunately, actual and/or simulated process fault data can also be obtained from other sources besides a high-fidelity dynamic simulation. Three such additional sources are as follows:

1. If the company developing the fault analyzer operates more than one of the target process systems, it is possible to pool the fault data collected from all these systems for use in the primary model testing. Westinghouse was able to use this technique effectively to gather test data when developing a diagnostic expert system for industrial turbines [16].

2. Another potential way to obtain process fault data is by altering normal process data to mimic process fault situations. To do this, the observed effects of a particular fault situation first need to be accurately predicted, and then the measured variables affected need to be altered accordingly in the normal process data collected. The measured variables being altered cannot be controlled variables in feedback control loops, trip variables for activating interlock shutdowns, and so on. If they are, the actual interaction between the various measured variables resulting from the fault situation might be too complex to predict accurately enough. This would be especially true for those process systems that are highly integrated or that produce multiple products.

3. A third way to obtain process fault data is to perform controlled experiments with the actual process system. These experiments should be relatively harmless and simple to perform. The results generate data for normal process operating events that also invalidate modeling assumptions in the primary models. During the FALCON project, DuPont performed many such experiments, and the data generated were used to test the fault analyzer.

If both steps of this verification procedure are performed correctly, the primary models will always be satisfied during normal process operations and will become violated whenever any of their underlying assumption variables deviate significantly. Models exhibiting such behavior will be referred to as being *well-formulated.* The biggest advantage of describing the process system as a set of well-formulated primary models is that the understanding of normal process behavior is represented by the summation of the knowledge contained within those models. This allows each primary model to be improved and maintained independent of the other primary models. The biggest advantages of having such independence within the declarative knowledge base are discussed further in Chapters 3 and 4. Briefly, with our automated MOME algorithm, it reduces the problem of achieving competent automating process fault analysis into the much simpler problem of process modeling, something most engineers better understand.

At this stage in the fault analyzer's development, the possible diagnostic information derived from the declarative knowledge still needs to be refined further into a SV&PFA diagnostic knowledge base. This is accomplished by creating SV&PFA diagnostic rules: patterns of diagnostic evidence capable of discriminating between the various possible process operating events, especially the various possible fault situations. However, as noted above, the procedures required to create these diagnostic rules are algorithmic in nature, so it was possible to automate the procedures to ensure that they are always applied properly. Consequently, the competency of the resulting fault analyzer depends entirely on the correctness of its declarative knowledge base, that is, the process knowledge contained within the various primary models. In turn, this depends entirely on how well the normal operating behavior of the target process system is understood and represented by the various primary models. *Thus, it is possible to guarantee that the fault analyzer will always perform competently if and only if all of the primary process models are guaranteed to be well-formulated.* This indicates that the main investment of effort during the development and verification of a model-based process fault analyzer should be in deriving and verifying the models and evaluating their corresponding required statistical parameters from sufficient normal process operating data.

2.5 SUMMARY

All sets of linearly independent models describing normal operation of the target process system to be monitored are classified as primary models. To be useful for sensor validation and proactive fault analysis, these models cannot rely on process variables that are either not measured directly or that cannot be adequately characterized by standard and/or extreme values of parameters under normal process operating conditions. The frequency with which the data are being collected must also be sufficient to properly evaluate time-dependent terms in the models (differential, integral, time delays, etc.). It is essential for correct performance by the fault analyzer to determine every modeling assumption variable on which these models depend and then determine the sensitivity of those models to the various possible modes of assumption variable deviations. Furthermore, the variance of each residual and its normal beta must be determined accurately as a function of process state (e.g., production rate). If done properly, the resulting primary models are termed well-formulated. Together, this understanding of normal process operations constitutes the deep knowledge (i.e., the declarative knowledge) that the KBS uses during its fault analysis. Creating this deep knowledge is thus accomplished using the following two-step procedure.

First, as many unique primary process models as possible should be derived to describe the normal operating behavior of the target process system. These models should be based on the most fundamental understanding of normal process behavior known and should be limited only by the specific type and frequency of process data being collected. The resulting set of models should constitute a highly detailed description of the target process system's normal operating behavior. All required modeling assumption variables must be included as appropriate terms in their respective primary models.

Second, it is necessary to determine how accurately these models actually predict a process system's normal operating behavior. Doing so requires a statistical analysis of the residuals resulting from an evaluation of these primary models with actual process data. We typically recommend using three to six months of faultless process data collected over a range (differing production rates, etc.) of normal operating conditions. This statistical analysis also requires correctly determining the sensitivity of those models with respect to each of their associated modeling assumption variables. Doing this properly determines the set of all possible secondary models. These two steps should ensure that the diagnostic evidence derived from the behavior of the models' residuals is always interpreted properly during subsequent fault analysis.

REFERENCES

1. Fickelscherer, R. J., *Automated Process Fault Analysis*, Ph.D. dissertation, University of Delaware, Newark, DE, 1990.
2. Venkatasubramanian, V., R. Rengaswamy, K. Yin, and S. N. Kavuri, "A Review of Process Fault Detection and Diagnosis: Part 1. Quantitative Model Based Methods," *Computers and Chemical Engineering*, Vol. 27, 2003, pp. 293–311.

3. Venkatasubramanian, V., R. Rengaswamy, and S. N. Kavuri, "A Review of Process Fault Detection and Diagnosis: Part 2. Qualitative Models and Search Strategies," *Computers and Chemical Engineering*, Vol. 27, 2003, pp. 313–326.

4. Venkatasubramanian, V., R. Rengaswamy, S. N. Kavuri, and K. Yin, "A Review of Process Fault Detection and Diagnosis: Part 3. Process History Based Methods," *Computers and Chemical Engineering*, Vol. 27, 2003, pp. 327–346.

5. Ma, J., and J. Jiang, "Applications of Fault Detection and Diagnosis Methods in Nuclear Power Plants: A Review," *Progress in Nuclear Energy*, Vol. 53, No. 3, 2011, pp. 255–266.

6. Fickelscherer, R. J., "A Generalized Approach to Model-Based Process Fault Analysis," in *Proceedings of the 2nd International Conference on Foundations of Computer-Aided Process Operations*, ed. by D. W. T. Rippin, J. C. Hale, and J. F. Davis, CACHE, Inc., Austin, TX, 1994, pp. 451–456.

7. Skotte, R., D. Lenz, R. Fickelscherer, W. An, D. Lapham III, C. Lymburner, J. Kaylor, D. Baptiste, M. Pinsky, F. Gani, and S. B. Jørgensen, "Advanced Processs Control with Innovation for an Integrated Electrochemical Process," presented at the *AIChE Spring National Meeting*, Houston, TX, 2001.

8. Fickelscherer, R. J., D. H. Lenz, and D. L. Chester, "Intelligent Process Supervision via Automated Data Validation and Fault Analysis: Results of Actual CPI Applications," Paper 115d, presented at the *AIChE Spring National Meeting*, New Orleans, LA, 2003.

9. Fickelscherer, R. J., D. H. Lenz, and D. L. Chester, "Fuzzy Logic Clarifies Operations," *InTech*, October 2005, pp. 53–57.

10. Zadeh, L. A., "Fuzzy Logic," *Computer*, Vol. 21, No. 4, 1988, pp. 83–93.

11. Kramer, M. A., and R. S. H. Mah, "Model-Based Monitoring," in *Foundations of Computer-Aided Process Operations II*, ed. by D. W. T. Rippin, J. C. Hale, and J. F. Davis, CACHE, Inc., Austin, TX, 1994, pp. 45–68.

12. Davis, R., and D. B. Lenat, *Knowledge-Based Systems in Artificial Intelligence*, McGraw-Hill, New York, 1982, p. 337.

13. Petti, T. F., *Using Mathematical Models in Knowledge Based Control Systems*, Ph.D. dissertation, Department of Chemical Engineering, University of Delaware, Newark, DE, 1989.

14. Kramer, M. A., Letter to the editor in regard to Petti et al., *AIChE Journal*, Vol. 36, 1990, p. 1121.

15. Koton, P. A., "Empirical and Model-Based Reasoning in Expert Systems," in *Proceedings of the Ninth International Joint Conference on Artificial Intelligence*, Los Angeles, Vol. 1, Morgan Kaufmann Publishers, Los Altos, CA, 1985, pp. 297–299.

16. Lowenfeld, S., and R. L. Osborne, "An Expert System for Process Diagnosis," in *Proceedings of the Fourth Annual Control Conference*, Rosemont, IL, 1984, pp. 19–28.

3

METHOD OF MINIMAL EVIDENCE: DIAGNOSTIC STRATEGY DETAILS

3.1 OVERVIEW

In this chapter we present a general methodology for creating optimal model-based process fault analyzers called the *method of minimal evidence* (MOME) [1, 2]. MOME is a diagnostic strategy based on an evaluation of engineering models describing normal operation of the target process system with sensor data. This methodology uses the minimum amount of diagnostic evidence necessary to discriminate uniquely between an invalid modeling assumption variable (e.g., an assumption that assumes the absence of a particular process fault situation) and all other valid modeling assumption variables. Moreover, it ensures that the resulting fault analyzer will always perform competently and optimizes the diagnostic sensitivity and resolution of its diagnoses. Diagnostic knowledge bases created with this methodology are also conducive for diagnosing many multiple-fault situations, for determining the strategic placement of process sensors to facilitate fault analysis, and for determining the shrewd distribution of fault analyzers within large processing plants. It has been demonstrated to be competent in both an adipic acid plant formerly owned and operated by DuPont in Victoria, Texas and an electrolytic persulfate plant owned and operated by FMC in Tonawanda, New York. The basic logical inference utilized by MOME is described in this chapter.

Optimal Automated Process Fault Analysis, First Edition.
Richard J. Fickelscherer and Daniel L. Chester.
© 2013 John Wiley & Sons, Inc. Published 2013 by John Wiley & Sons, Inc.

3.2 INTRODUCTION

As described in Chapter 2, diagnostic evidence is inferred from the residuals of evaluated primary and secondary models. The values of the residuals indicate which models are satisfied and which are violated. The violation of a well-formulated primary model implies that at least one of its associated modeling assumption variables that can cause such a violation is invalid. In contrast, the satisfaction of a well-formulated primary model implies either that all of its associated modeling assumption variables are valid (i.e., at least not deviating significantly for that model), or that two or more of those assumption variables are deviating significantly but are interacting in such a way as to cause that model to appear satisfied.

Once the primary and secondary models have been evaluated, plausible hypotheses as to the underlying cause or causes of the abnormal process behavior observed can be derived by logically interpreting the resulting patterns of model satisfactions and violations. Fault analysis performed in this manner is said to follow a model-based diagnostic strategy. Currently, a wide variety of model-based diagnostic strategies exist [3, 4].[1] However, none of these strategies optimize the diagnostic sensitivity and resolution of diagnostic rules they create while guaranteeing that the resulting fault analyzer will always perform competently. A model-based diagnostic strategy that does so, the method of minimum evidence, is described here.

The MOME diagnostic strategy for fault analysis compares patterns of residual behavior expected to occur during the various possible assumption variable deviations with the patterns of those residuals currently present in the process. Any matches identify the underlying possible assumption variable(s) deviations currently occurring in the process. MOME uses the minimum unique patterns required to carryout this analysis correctly, making possible direct identification of many of the possible multiple assumption deviations. This is important because this methodology discerns not only process faults but also all possible process operating events, both fault and nonfault events. The logic is designed to venture unique diagnoses only when there is high certainty of the underlying problem. As discussed in Appendix B, such conservative behavior is advantageous because it does not confuse its users with logically implausible diagnoses at times when the actual process operating state is in flux.

The methodology is based on *default reasoning*. All but one (if perfect resolution between the various potential assumption deviations is possible from the current evidence) of the various potential assumption variable deviations

---

[1]A number of various model-based diagnostic strategies are also cited in Section A.7.

for a violated model are shown not to be deviating by other violated and satisfied models in the expected pattern of residual behavior. Consequently, the most plausible explanation for such a situation is then the remaining assumption variable deviation(s).

The patterns of expected residual behavior resulting from applying this method [e.g., the sensor validation and proactive fault analysis (SV&PFA) diagnostic rules] contain the minimum patterns required to diagnose each of the possible fault situations. This maximizes the sensitivity of the fault analyzer for these various faults, maximizes the resolution (discrimination between various possible fault and nonfault events for the magnitude of fault occurring) of that analysis, and optimizes its overall competence when confronted with multiple assumption deviations. The methodology can be used to determine the strategic placement of process sensors for performing fault analysis and the shrewd division of large process systems for distributing process fault analyzers.

3.3 MOME DIAGNOSTIC STRATEGY

Our model-based diagnostic strategy for automating process fault analysis is described next.

3.3.1 Example of MOME SV&PFA Diagnostic Rules' Logic

Example 3.1   Consider further the instrumented process mixing tee shown in Figure 2.5 and modeled in detail in Example 2.2. The following example of the MOME diagnostic strategy demonstrates some of the patterns of expected model residuals (i.e., SV&PFA diagnostic rules) from which the current operating state of this mixing tee process can be inferred.[2]

The various possible combinations of residual behavior that can occur whenever any of the nine linear modeling assumption variables used to derive primary models P_1 and P_2 and secondary models S_1, S_2, and S_3 (Chapter 2) deviate significantly can be predetermined and written down as SV&PFA diagnostic rules to identify those invalid assumption variables. To derive a diagnosis of the current process state, the pattern of residual behavior determined from current sensor data can then be compared against the patterns

---

[2]These engineering models are contained in the SV&PFA module of the **FALCONEER**™ **IV** application currently running online at FMC's Tonawanda electrolytic sodium persulfate plant. **FALCONEER** is an acronym for "fault analysis consultant via engineering equation residuals."

expected. The nomenclature and logic symbols used in the diagnostic rules are defined below.[3]

MOME Diagnostic Rule Nomenclature

$$P_X^Y : \text{primary model } X \text{ residual is in state } Y$$

where state Y can be qualified as follows:

H = the residual is considered significantly **high**
S = the residual is considered neither significantly high nor significantly low (e.g., it is normal or, in other words, **satisfied**)
L = the residual is considered significantly **low**

$$S_X^Y: \text{secondary model } X \text{ residual is in state } Y$$

where state Y can be qualified as follows:

H = the residual is considered significantly **high**
S = the residual is considered neither significantly high nor significantly low (e.g., it is normal or, in other words, **satisfied**)
L = the residual is considered significantly **low**

$$a_X^Y: \text{modeling assumption variable } X \text{ is in state } Y^{\,4}$$

where state Y can be qualified as follows:

H = the assumption variable is deviating significantly **high**
S = the assumption variable is neither deviating significantly high nor significantly low (e.g., it is **satisfied**)
L = the assumption variable is deviating significantly **low**

---

[3]The following analysis and subsequent discussion are based on Boolean logic. This does not limit its generality in any way. This analytical method is formalized further and expanded to non-Boolean reasoning in Chapter 4, specifically to rely on fuzzy logic reasoning and certainty factor calculations.

[4]The value of state Y for a given assumption variable depends on whether that variable is for the accuracy of either a sensor measurement or a constant parameter in the relevant primary and secondary models. This state for assumption deviations caused by inaccurate sensor measurements which violate models that depend on them is in the same direction as the signs of those models' first derivatives with respect to those assumptions. This state for assumption deviations caused by incorrect values of constant parameters is just the opposite of those signs. Such assignments are demonstrated in this example.

Logic Symbols

∧ represents conjunction: logical AND

∨ represents disjunction: logical inclusive OR

⊖ represents disjunction: logical exclusive OR

¬ represents negation: logical NOT

→ represents implication: logical implication

↔ represents equivalence: logical equivalence

The residuals that result from evaluating the primary and secondary models with current process data are all judged to be either low, satisfied, or high states. High and low residual states indicate that at least one assumption variable that can cause such behavior is deviating significantly. Satisfied residual states indicate either (1) that there are no modeling assumption variable deviations; (2) that one or more such deviations are occurring but at magnitudes or rates of change below the sensitivity of that model to discriminate such deviations (they are not significant to that model), or (3) that two or more significant assumption variable deviations are interacting in an opposing fashion. Standard logical inference can be used directly on the resulting residual patterns to identify the set of violated and satisfied assumption variables.

Consider the situation where all the residuals of the two primary and three secondary models above are satisfied.[5] For single-fault situations, this would normally result in the inference that all nine of the associated assumption variables are satisfied (i.e., not deviating sufficiently to violate any of the models). The smallest value of the given assumption deviation that will violate at least one of these two primary models directly defines the minimum sensitivity of the resulting fault analyzer to each of the nine modeling assumption variables above. As discussed in Chapter 2, such deviations are defined as minimum assumption variable deviations [equation (2.9)] and are typically different

---

[5]All five models are considered *relevant* in the following analysis for all 15 possible assumption variable deviations. The MOME methodology does its analysis only on relevant primary and secondary models. All primary and secondary models directly dependent on a given assumption variable are considered relevant to that possible assumption variable deviation, as are all secondary models formed by eliminating the given assumption variable. So are any primary models that are required to create secondary models dependent on the given assumption variable but that do not require that assumption variable themselves. The set of all other possible primary and secondary models used to describe normal operating behavior not included in the relevant set are consequently considered irrelevant in our analysis for deviations of that given modeling assumption variable.

Table 3.1 $\partial \varepsilon_i / \partial a_j$

Model	P_1	P_2	S_1	S_2	S_3
Assumption variable					
a_1 sensor	$+500$	$+900T1$	$T1 - T2$	0	$T1 - T3$
a_2 sensor	$+500$	$+900T2$	0	$T2 - T1$	$T2 - T3$
a_3 sensor	-500	$-900T3$	$T2 - T3$	$T1 - T3$	0
a_4 parameter	-500	$-900T1$	$T2 - T1$	0	$T3 - T1$
a_5 parameter	-500	$-900T2$	0	$T1 - T2$	$T3 - T2$
a_6 parameter	-500	$-900T3$	$T2 - T3$	$T1 - T3$	0
a_7 sensor	0	$+900(F1 - L1)$	$F1 - L1$	$F3 + L3 - F2 + L2$	$F1 - L1$
a_8 sensor	0	$+900(F2 - L2)$	$F3 + L3 - F1 + L1$	$F2 - L2$	$F2 - L2$
a_9 sensor	0	$-900(F3 + L3)$	$-F3 - L3$	$-F3 - L3$	$L2 - F2 - F1 + L1$

for each primary and secondary model. The corresponding sensor validation diagnostic rule would thus be as follows:

Sensor Validation Diagnostic Rule

$$P_1{}^S \wedge P_2{}^S \wedge S_1{}^S \wedge S_2{}^S \wedge S_3{}^S \rightarrow a_1{}^S \wedge a_2{}^S \wedge a_3{}^S \wedge a_4{}^S \wedge a_5{}^S \wedge a_6{}^S \wedge a_7{}^S \wedge a_8{}^S \wedge a_9{}^S$$

The patterns of violated and satisfied residuals (i.e., the other possible SV&PFA diagnostic rules) that would be expected when an assumption variable deviated significantly can easily be constructed for this instrumented process mixing tee. The diagnostic rules can be created systematically by the following MOME procedure. First, the states of the violated residuals when particular significant assumption variable deviations occur are determined. This is accomplished by initially evaluating the sign of the first derivative of each model to each assumption variable. These first derivatives (i.e., $\partial \varepsilon_i / \partial a_j$) are listed in Table 3.1.

For argument's sake, in this example assume that flow rate F1 is greater than flow rate F2 (i.e., F1 > F2) and temperature T1 is greater than T2 (i.e., T1 > T2). These assumptions are arbitrary but are required in order to evaluate the first derivatives and they do not affect the generality of the following discussion.[6] These assumptions lead directly to Table 3.2, the signs of these derivatives [i.e., sign($\partial \varepsilon_i / \partial a_j$)].[7]

---

[6]From the physics of this unit operation, these two assumptions mandate that F3 > F1 > F2 and T1 > T3 > T2.

[7]FALCONEER™ IV computes these signs with current process data on each analysis cycle and adjusts its inferencing accordingly.

This determination of signs is required for the following reason. If the sign of this derivative is positive, the associated model residual will go high for significantly high assumption deviations of measured variables (sensors above) and go low for significantly high assumption deviations of unmeasured variables (parameters above). Similarly, this residual will go low and high, respectively, for the case of significantly low assumption deviations. Just the opposite is true if the sign of this derivative is negative. If the derivative evaluates to zero, the associated residual is unaffected by the indicated assumption variable deviations. To summarize this *assignment logic:*

```
IF
sign(∂ε_i/∂a_j)      AND a_j
IS                   IS A            THEN
        +            sensor          a_j^H → ε_i^S ∨ ε_i^H   &   a_j^L → ε_i^S ∨ ε_i^L
        +            parameter       a_j^H → ε_i^S ∨ ε_i^L   &   a_j^L → ε_i^S ∨ ε_i^H
        -            sensor          a_j^H → ε_i^S ∨ ε_i^L   &   a_j^L → ε_i^S ∨ ε_i^H
        -            parameter       a_j^H → ε_i^S ∨ ε_i^H   &   a_j^L → ε_i^S ∨ ε_i^L
        0            sensor          a_j^H → ε_i^S           &   a_j^L → ε_i^S
        0            parameter       a_j^H → ε_i^S           &   a_j^L → ε_i^S
```

Before diagnostic patterns can be created, one more issue needs to be examined. The parameters required to derive the primary models are set to their normal and/or extreme values when those models are evaluated to generate their corresponding residuals. If it is truly an extreme value (in this example, leaks are assumed not to be occurring, so their normal and extreme value would be 0.0), the actual value of the parameter can deviate in one direction only (for the leak this would be only positive values if the system is pressurized and be only negative values if the system is under vacuum). For this example, the three possible leaks can deviate high and significant values would cause the two primary models to be violated high [for $sign(\partial \varepsilon_i / \partial a_j)$ negative, $a_j^H \to \varepsilon_i^S \vee \varepsilon_i^H$ and a_j^L does not occur and thus can be ignored in the following analysis].

As is obvious from this assignment logic, the various relationships are describing cause (assumption variable deviation) to effect (residual response). In fault diagnosis, we actually reason from observed effects to underlying causes, a reasoning mechanism known as *abduction* [5]. Succinctly, if A implies B and B is true, we then conclude that A is true. This is called *causality* and is not the same as logical implication.[8] Consequently, abduction is only plausible inference: There could be many other plausible explanations

---

[8]In logical implication, knowing A and knowing A implies B, we then know B. This rule of inference is usually referred to as a *syllogism* [6].

Table 3.2 $\text{sign}(\partial \varepsilon_i / \partial a_j)$

Model	P_1	P_2	S_1	S_2	S_3
Assumption variable					
a_1 sensor	+	+	+	0	+
a_2 sensor	+	+	0	−	−
a_3 sensor	−	−	−	+	0
a_4 parameter	−	−	−	0	−
a_5 parameter	−	−	0	+	+
a_6 parameter	−	−	−	+	0
a_7 sensor	0	+	+	+	+
a_8 sensor	0	+	+	+	+
a_9 sensor	0	−	−	−	−

for B being true. A technique commonly used to reduce the other plausible explanations as much as possible is the *closed-world assumption* [7, 8] (regarded as the most straightforward and simplest way to complete a theory). This reduces the possible set of all potentially plausible hypotheses only to those referred to explicitly in the causality logic.

The sign table (Table 3.2), together with the assignment logic above, is the basis for performing *pattern matching*, an area widely studied in the field of artificial intelligence [9]. Pattern matching is a process of assigning values to unbound variables consistent with the constraints imposed by the relevant facts and rules [10]. The patterns useful for direct proactive fault analysis are derived by the following set-covering technique. Each model can fail either high or low or can be satisfied. In single-fault situations, the fact that a model is satisfied does not imply that all its associated assumption deviations are satisfied, only that the magnitude of the given assumption deviation is insufficient to violate the associated models (Table 3.3). The patterns of residual response identified by the assignment logic above superimposed on the sign table creates the possibility of many potential fault patterns.

Unique patterns in the various rows in Table 3.3 constitute valid diagnostic rules that offer perfect resolution of their associated assumption variable deviations. A rule may be found for each assumption variable deviation by choosing as many residual response deviations as possible from those appearing in that assumption variable deviation's row and adding satisfied responses for the other relevant residuals. If other assumption variable deviations are compatible with this pattern, the associated diagnostic rule has to conclude a disjunction of all those assumption deviations, conjoined with the satisfied state of all the assumptions whose deviations are not compatible. When more than one assumption variable deviation is compatible with the pattern, the

Table 3.3 a_j Deviation Consequenses

Model	P_1	P_2	S_1	S_2	S_3
Assumption Deviation					
$a_1{}^L$	$P_1{}^S \vee P_1{}^L$	$P_2{}^S \vee P_2{}^L$	$S_1{}^S \vee S_1{}^L$	$S_2{}^S$	$S_3{}^S \vee S_3{}^L$
$a_1{}^H$	$P_1{}^S \vee P_1{}^H$	$P_2{}^S \vee P_2{}^H$	$S_1{}^S \vee S_1{}^H$	$S_2{}^S$	$S_3{}^S \vee S_3{}^H$
$a_2{}^L$	$P_1{}^S \vee P_1{}^L$	$P_2{}^S \vee P_2{}^L$	$S_1{}^S$	$S_2{}^S \vee S_2{}^H$	$S_3{}^S \vee S_3{}^H$
$a_2{}^H$	$P_1{}^S \vee P_1{}^H$	$P_2{}^S \vee P_2{}^H$	$S_1{}^S$	$S_2{}^S \vee S_2{}^L$	$S_3{}^S \vee S_3{}^L$
$a_3{}^L$	$P_1{}^S \vee P_1{}^H$	$P_2{}^S \vee P_2{}^H$	$S_1{}^S \vee S_1{}^H$	$S_2{}^S \vee S_2{}^L$	$S_3{}^S$
$a_3{}^H$	$P_1{}^S \vee P_1{}^L$	$P_2{}^S \vee P_2{}^L$	$S_1{}^S \vee S_1{}^L$	$S_2{}^S \vee S_2{}^H$	$S_3{}^S$
$a_4{}^H$	$P_1{}^S \vee P_1{}^H$	$P_2{}^S \vee P_2{}^H$	$S_1{}^S \vee S_1{}^H$	$S_2{}^S$	$S_3{}^S \vee S_3{}^H$
$a_5{}^H$	$P_1{}^S \vee P_1{}^H$	$P_2{}^S \vee P_2{}^H$	$S_1{}^S$	$S_2{}^S \vee S_2{}^L$	$S_3{}^S \vee S_3{}^L$
$a_6{}^H$	$P_1{}^S \vee P_1{}^H$	$P_2{}^S \vee P_2{}^H$	$S_1{}^S \vee S_1{}^H$	$S_2{}^S \vee S_2{}^L$	$S_3{}^S$
$a_7{}^L$	$P_1{}^S$	$P_2{}^S \vee P_2{}^L$	$S_1{}^S \vee S_1{}^L$	$S_2{}^S \vee S_2{}^L$	$S_3{}^S \vee S_3{}^L$
$a_7{}^H$	$P_1{}^S$	$P_2{}^S \vee P_2{}^H$	$S_1{}^S \vee S_1{}^H$	$S_2{}^S \vee S_2{}^H$	$S_3{}^S \vee S_3{}^H$
$a_8{}^L$	$P_1{}^S$	$P_2{}^S \vee P_2{}^L$	$S_1{}^S \vee S_1{}^L$	$S_2{}^S \vee S_2{}^L$	$S_3{}^S \vee S_3{}^L$
$a_8{}^H$	$P_1{}^S$	$P_2{}^S \vee P_2{}^H$	$S_1{}^S \vee S_1{}^H$	$S_2{}^S \vee S_2{}^H$	$S_3{}^S \vee S_3{}^H$
$a_9{}^L$	$P_1{}^S$	$P_2{}^S \vee P_2{}^H$	$S_1{}^S \vee S_1{}^H$	$S_2{}^S \vee S_2{}^H$	$S_3{}^S \vee S_3{}^H$
$a_9{}^H$	$P_1{}^S$	$P_2{}^S \vee P_2{}^L$	$S_1{}^S \vee S_1{}^L$	$S_2{}^S \vee S_2{}^L$	$S_3{}^S \vee S_3{}^L$

associated diagnosis has lower diagnostic resolution. Sometimes the pattern can be relaxed by replacing a residual response deviation with the disjunction of that deviation with that residual's satisfied response without changing the set of assumption variables with which the pattern is compatible (i.e., diagnostic resolution is not diminished with this disjunction). The following SV&PFA diagnostic rules can be derived from this table to identify the various associated fault situations with the highest diagnostic resolution.

Sensor Validation and Proactive Fault Analysis Diagnostic Rules   At the highest possible levels of diagnostic resolution (i.e., discrimination possible between different faults):

$$P_1{}^L \wedge \left(P_2{}^L \vee P_2{}^S\right) \wedge S_1{}^L \wedge S_2{}^S \wedge S_3{}^L$$
$$\rightarrow a_1{}^L \wedge a_2{}^S \wedge a_3{}^S \wedge a_4{}^S \wedge a_5{}^S \wedge a_6{}^S \wedge a_7{}^S \wedge a_8{}^S \wedge a_9{}^S$$

$$P_1{}^L \wedge \left(P_2{}^L \vee P_2{}^S\right) \wedge S_1{}^S \wedge S_2{}^H \wedge S_3{}^H$$
$$\rightarrow a_2{}^L \wedge a_1{}^S \wedge a_3{}^S \wedge a_4{}^S \wedge a_5{}^S \wedge a_6{}^S \wedge a_7{}^S \wedge a_8{}^S \wedge a_9{}^S$$

$$P_1{}^S \wedge P_2{}^L \wedge S_1{}^S \wedge \left(S_2{}^S \vee S_2{}^H\right) \wedge S_3{}^H$$
$$\rightarrow a_2{}^L \wedge a_1{}^S \wedge a_3{}^S \wedge a_4{}^S \wedge a_5{}^S \wedge a_6{}^S \wedge a_7{}^S \wedge a_8{}^S \wedge a_9{}^S$$

$$P_1{}^L \wedge \left(P_2{}^L \vee P_2{}^S\right) \wedge S_1{}^L \wedge S_2{}^H \wedge S_3{}^S$$
$$\rightarrow a_3{}^H \wedge a_1{}^S \wedge a_2{}^S \wedge a_4{}^S \wedge a_5{}^S \wedge a_6{}^S \wedge a_7{}^S \wedge a_8{}^S \wedge a_9{}^S$$

$$P_1{}^S \wedge P_2{}^L \wedge S_1{}^L \wedge S_2{}^H \wedge S_3{}^S$$
$$\rightarrow a_3{}^H \wedge a_1{}^S \wedge a_2{}^S \wedge a_4{}^S \wedge a_5{}^S \wedge a_6{}^S \wedge a_7{}^S \wedge a_8{}^S \wedge a_9{}^S$$

$$P_1{}^H \wedge \left(P_2{}^H \vee P_2{}^S\right) \wedge S_1{}^H \wedge S_2{}^S \wedge S_3{}^H$$
$$\rightarrow \left(a_1{}^H \vee a_4{}^H\right) \wedge a_2{}^S \wedge a_3{}^S \wedge a_5{}^S \wedge a_6{}^S \wedge a_7{}^S \wedge a_8{}^S \wedge a_9{}^S$$

$$P_1{}^H \wedge \left(P_2{}^H \vee P_2{}^S\right) \wedge S_1{}^S \wedge S_2{}^L \wedge S_3{}^L$$
$$\rightarrow \left(a_2{}^H \vee a_5{}^H\right) \wedge a_1{}^S \wedge a_3{}^S \wedge a_4{}^S \wedge a_6{}^S \wedge a_7{}^S \wedge a_8{}^S \wedge a_9{}^S$$

$$P_1{}^S \wedge P_2{}^H \wedge S_1{}^S \wedge \left(S_2{}^L \vee S_2{}^S\right) \wedge S_3{}^L$$
$$\rightarrow \left(a_1{}^H \vee a_5{}^H\right) \wedge a_1{}^S \wedge a_3{}^S \wedge a_4{}^S \wedge a_6{}^S \wedge a_7{}^S \wedge a_8{}^S \wedge a_9{}^S$$

$$P_1{}^H \wedge \left(P_2{}^H \vee P_2{}^S\right) \wedge S_1{}^H \wedge S_2{}^L \wedge S_3{}^S$$
$$\rightarrow \left(a_3{}^L \vee a_6{}^H\right) \wedge a_1{}^S \wedge a_2{}^S \wedge a_4{}^S \wedge a_5{}^S \wedge a_7{}^S \wedge a_8{}^S \wedge a_9{}^S$$

$$P_1{}^S \wedge P_2{}^H \wedge S_1{}^H \wedge S_2{}^L \wedge S_3{}^S$$
$$\rightarrow \left(a_3{}^L \vee a_6{}^H\right) \wedge a_1{}^S \wedge a_2{}^S \wedge a_4{}^S \wedge a_5{}^S \wedge a_7{}^S \wedge a_8{}^S \wedge a_9{}^S$$

$$P_1{}^S \wedge P_2{}^L \wedge \left(S_1{}^L \vee S_1{}^S\right) \wedge S_2{}^L \wedge \left(S_3{}^L \vee S_3{}^S\right)$$
$$\rightarrow \left(a_7{}^L \vee a_8{}^L \vee a_9{}^H\right) \wedge a_1{}^S \wedge a_2{}^S \wedge a_3{}^S \wedge a_4{}^S \wedge a_5{}^S \wedge a_6{}^S$$

$$P_1{}^S \wedge P_2{}^H \wedge \left(S_1{}^H \vee S_1{}^S\right) \wedge S_2{}^H \wedge \left(S_3{}^H \vee S_3{}^S\right)$$
$$\rightarrow \left(a_7{}^H \vee a_8{}^H \vee a_9{}^L\right) \wedge a_1{}^S \wedge a_2{}^S \wedge a_3{}^S \wedge a_4{}^S \wedge a_5{}^S \wedge a_6{}^S$$

These complete patterns do not always develop if the magnitudes of the assumption variable deviations occur at levels that violate some but not all model residuals (i.e., as discussed in Chapter 2, these theoretical minimum assumption deviations [determined using equation (2.9)] are normally different for each model. By relaxing these complete patterns further, new assumption variable deviations become compatible with them and have to be added as part of a disjunction of deviations in the conclusions of the rules. This leads to more ambiguous diagnoses (i.e., lower diagnostic resolution) as demonstrated by the following SV&PFA diagnostic rules.[9]

[9]The original FALCON system included only those diagnostic rules for diagnosing faults at the highest possible level of diagnostic resolution.

Sensor Validation and Proactive Fault Analysis Diagnostic Rules At
lower possible levels of diagnostic resolution but higher levels of diagnostic
sensitivity:

$$P_1{}^H \wedge \left(P_2{}^H \vee P_2{}^S\right) \wedge S_1{}^H \wedge S_2{}^S \wedge S_3{}^S$$
$$\rightarrow \left(a_1{}^H \vee a_3{}^L \vee a_4{}^H \vee a_6{}^H\right) \wedge a_2{}^S \wedge a_5{}^S \wedge a_7{}^S \wedge a_8{}^S \wedge a_9{}^S$$

$$P_1{}^H \wedge \left(P_2{}^H \vee P_2{}^S\right) \wedge S_1{}^S \wedge S_2{}^L \wedge S_3{}^S$$
$$\rightarrow \left(a_2{}^H \vee a_3{}^L \vee a_5{}^H \vee a_6{}^H\right) \wedge a_1{}^S \wedge a_4{}^S \wedge a_7{}^S \wedge a_8{}^S \wedge a_9{}^S$$

$$P_1{}^L \wedge \left(P_2{}^L \vee P_2{}^S\right) \wedge S_1{}^L \wedge S_2{}^S \wedge S_3{}^S$$
$$\rightarrow \left(a_1{}^L \vee a_3{}^H\right) \wedge a_2{}^S \wedge a_4{}^S \wedge a_5{}^S \wedge a_6{}^S \wedge a_7{}^S \wedge a_8{}^S \wedge a_9{}^S$$

$$P_1{}^L \wedge \left(P_2{}^L \vee P_2{}^S\right) \wedge S_1{}^S \wedge S_2{}^H \wedge S_3{}^S$$
$$\rightarrow \left(a_2{}^L \vee a_3{}^H\right) \wedge a_1{}^S \wedge a_4{}^S \wedge a_5{}^S \wedge a_6{}^S \wedge a_7{}^S \wedge a_8{}^S \wedge a_9{}^S$$

$$P_1{}^S \wedge P_2{}^H \wedge S_1{}^H \wedge S_2{}^S \wedge S_3{}^H$$
$$\rightarrow \left(a_1{}^H \vee a_4{}^H \vee a_7{}^H \vee a_8{}^H \vee a_9{}^L\right) \wedge a_2{}^S \wedge a_3{}^S \wedge a_5{}^S \wedge a_6{}^S$$

$$P_1{}^S \wedge P_2{}^H \wedge S_1{}^S \wedge S_2{}^L \wedge S_3{}^S$$
$$\rightarrow \left(a_2{}^H \vee a_3{}^L \vee a_5{}^H \vee a_6{}^H\right) \wedge a_1{}^S \wedge a_4{}^S \wedge a_7{}^S \wedge a_8{}^S \wedge a_9{}^S$$

$$P_1{}^S \wedge P_2{}^H \wedge S_1{}^H \wedge S_2{}^S \wedge S_3{}^S$$
$$\rightarrow \left(a_1{}^H \vee a_3{}^L \vee a_4{}^H \vee a_6{}^H \vee a_7{}^H \vee a_8{}^H \vee a_9{}^L\right) \wedge a_2{}^S \wedge a_5{}^S$$

$$P_1{}^S \wedge P_2{}^H \wedge S_1{}^S \wedge S_2{}^S \wedge S_3{}^H$$
$$\rightarrow \left(a_1{}^H \vee a_4{}^H \vee a_7{}^H \vee a_8{}^H \vee a_9{}^L\right) \wedge a_2{}^S \wedge a_3{}^S \wedge a_5{}^S \wedge a_6{}^S$$

$$P_1{}^S \wedge P_2{}^L \wedge S_1{}^L \wedge S_2{}^S \wedge S_3{}^L$$
$$\rightarrow \left(a_1{}^L \vee a_7{}^L \vee a_8{}^L \vee a_9{}^H\right) \wedge a_2{}^S \wedge a_3{}^S \wedge a_4{}^S \wedge a_5{}^S \wedge a_6{}^S$$

$$P_1{}^S \wedge P_2{}^L \wedge S_1{}^S \wedge S_2{}^H \wedge S_3{}^S$$
$$\rightarrow \left(a_2{}^L \vee a_3{}^H\right) \wedge a_1{}^S \wedge a_4{}^S \wedge a_5{}^S \wedge a_6{}^S \wedge a_7{}^S \wedge a_8{}^S \wedge a_9{}^S$$

$$P_1{}^S \wedge P_2{}^L \wedge S_1{}^L \wedge S_2{}^S \wedge S_3{}^S$$
$$\rightarrow \left(a_1{}^L \vee a_3{}^H \vee a_7{}^L \vee a_8{}^L \vee a_9{}^H\right) \wedge a_2{}^S \wedge a_4{}^S \wedge a_5{}^S \wedge a_6{}^S$$

$$P_1{}^S \wedge P_2{}^L \wedge S_1{}^S \wedge S_2{}^S \wedge S_3{}^L$$
$$\rightarrow \left(a_1{}^L \vee a_7{}^L \vee a_8{}^L \vee a_9{}^H\right) \wedge a_2{}^S \wedge a_3{}^S \wedge a_4{}^S \wedge a_5{}^S \wedge a_6{}^S$$

Early fault detection and diagnosis by MOME requires only that one primary model be violated. This results in the most sensitive diagnosis possible but also in the lowest diagnostic resolution between fault hypotheses by the fault analyzer. This trades diagnostic resolution directly for the earliest possible alert for this model-based analysis.

Sensor Validation and Proactive Fault Analysis Diagnostic Rules At the lowest possible levels of diagnostic resolution but the highest possible levels of diagnostic sensitivity:

$$P_1^L \wedge \left(P_2^L \vee P_2^S\right) \wedge S_1^S \wedge S_2^S \wedge S_3^S$$
$$\rightarrow \left(a_1^L \vee a_2^L \vee a_3^H\right) \wedge a_4^S \wedge a_5^S \wedge a_6^S \wedge a_7^S \wedge a_8^S \wedge a_9^S$$

$$P_1^H \wedge \left(P_2^H \vee P_2^S\right) \wedge S_1^S \wedge S_2^S \wedge S_3^S$$
$$\rightarrow \left(a_1^H \vee a_2^H \vee a_3^L \vee a_4^H \vee a_5^H \vee a_6^H\right) \wedge a_7^S \wedge a_8^S \wedge a_9^S$$

$$P_1^S \wedge P_2^H \wedge S_1^S \wedge S_2^S \wedge S_3^S$$
$$\rightarrow \left(a_1^H \vee a_2^H \vee a_3^L \vee a_4^H \vee a_5^H \vee a_6^H \vee a_7^H \vee a_8^H \vee a_9^L\right)$$

$$P_1^S \wedge P_2^L \wedge S_1^S \wedge S_2^S \wedge S_3^S$$
$$\rightarrow \left(a_1^L \vee a_2^L \vee a_3^H \vee a_7^L \vee a_8^L \vee a_9^H\right) \wedge a_4^S \wedge a_5^S \wedge a_6^S$$

As demonstrated in this example, perfect resolution between different process fault hypotheses is not always possible or is possible only at larger magnitudes of the fault specified (i.e., at a magnitude sufficient to violate all relevant primary and secondary model residuals so affected). This is the classic trade-off between timely fault detection and correct identification of the underlying fault(s). Trading lower diagnostic resolution for higher diagnostic sensitivity in the fashion demonstrated above allows the fault analyzer to narrow the potential process faults that could be occurring currently into a reasonable number of plausible explanations for the current process state that can then be checked out further by the process operator to determine the actual fault present. This flags potential incipient fault situations sooner rather than waiting until the fault's magnitude is severe enough to allow unique classification. This is why the methodology is called the method of minimal evidence (MOME): All plausible fault situations are diagnosed whenever even only just one primary model residual indicates abnormal process operation.

It is important to emphasize that because of the extensive use of inclusive OR logic in these rules, whenever more than one fault hypothesis is announced, it could actually be that one of the others or all is the actual fault situation occurring in the process (i.e., all hypotheses are considered equally plausible). In those cases, additional reasoning (i.e., based on meta-knowledge

[11]) is required to resolve such ambiguous situations. Furthermore, because the MOME algorithm normally also cross-checks the current status of the states of all possible relevant primary and secondary models' residuals for each potential assumption deviation, as many such inclusive OR logic ambiguities as possible are eliminated. A generalized fuzzy logic diagnostic rule implanting the MOME strategy for single-fault situations, currently implemented in FALCONEER^{TM} IV, is described in detail in Chapter 4.

Finally, almost any other combination of primary and secondary models both violated and satisfied not shown above would indicate that a multiple-fault situation was occurring. There are 243 ($= 3^5$) possible SV&PFA diagnostic rules for this process system. The example above shows only 51 of these possible patterns.[10] These 51 patterns and most of the remaining 192 possible patterns are also interpreted by another generalized fuzzy logic diagnostic rule also currently implemented in FALCONEER^{TM} IV that diagnoses multiple faults directly. This rule is also described in detail in Chapter 4.

It should now be highly evident that automating the creation of this diagnostic logic is extremely advantageous. Even for a process system as trivial as this mixing tee requires a substantial effort to analyze a mere handful (15 possible assumption variable deviations) of potential process faults. When there are scores of primary models and hundreds of necessary assumption variables, literally tens of thousands of possible SV&PFA diagnostic rules are required for competent analysis of the current process operating state.

3.3.2 Example of Key Performance Indicator Validation

Example 3.2   Consider again the heat exchanger described in Example 2.1. If the energy balance derived there (primary model $\mathbf{P_0}^*$) was the only primary model depending on each of its entire set of assumptions, the unique diagnostic rule set for this system would be as follows:

$$\mathbf{P_0}^*: \quad \varepsilon_{P0}^* = (F_{\text{Pi}} - L_{\text{Pup}})c_{P} T_{\text{Pi}} - (F_{\text{Pi}} + L_{\text{Pdn}})c_{P} T_{\text{Po}}$$

$$- \exp(P_{\text{FAIL}}) \left(\frac{P_{\text{eff}}}{100.0} \right) \ln(L_{\text{TUBES}}) \left[(F_{\text{Wex}} + L_{\text{Wdn}})c_{W} T_{\text{Wo}} \right.$$

$$\left. - (F_{\text{Wex}} - L_{\text{Wup}})c_{W} T_{\text{Wi}}) \right]$$

$$- M_{\text{Wex}} c_{W} \frac{d(T_{\text{Wo}})}{d(\text{time})}$$

---

[10]It is possible to derive other patterns for diagnosing single-fault situations in this example that do not rely on either primary model being violated, but these are currently suppressed in FALCONEER^{TM} IV. We require at least one primary model to be violated for any fault diagnosis. In actual applications this makes the results more meaningful, although in some odd situations not as sensitive as theoretically possible. See Example 3.4 for a further explanation about this exclusion.

Assumption variables for primary model $\mathbf{P_0}^*$:

a_{10}:   Linear; F_{Pi} (process stream flow rate into HTX) flow sensor is correct

a_{11}:   Linear; T_{Pi} (process stream temperature into HTX) thermocouple reading is correct

a_{12}:   Linear; T_{Po} (process stream temperature out of HTX) thermocouple reading is correct

a_{13}:   Linear; T_{Wi} (water stream temperature into HTX) thermocouple reading is correct

a_{14}:   Linear; T_{Wo} (water stream temperature out of HTX) thermocouple reading is correct

a_{15}:   Linear; cp_P (heat capacity of process stream) is known and constant

a_{16}:   Linear; cp_W (heat capacity of water stream) is known and constant

a_{17}:   Linear; M_{Wex} (mass of water in shell side of HTX) is known and constant

a_{18}:   Linear; L_{Pup}— no process leaks upstream from HTX

a_{19}:   Linear; L_{Pdn}— no process leaks downstream from HTX

a_{20}:   Continuous nonlinear; L_{TUBES}—no water leaks in HTX tubes

a_{21}:   Linear; L_{Wup}—no cooling water leaks upstream from HTX

a_{22}:   Linear; L_{Wdn}—no cooling water leaks downstream from HTX

a_{23}:   Linear; P_{eff}—no gradual decrease in the cooling water pump efficiency

a_{24}:   Linear; F_{Wex} (flow of water in shell side of HTX) is known and constant

a_{25}:   Discrete nonlinear; P_{FAIL} no sudden, complete failure of the cooling-water pump

Next we describe the three possible SV&PFA rules when having only one primary model.

Sensor Validation Diagnostic Rule

$$P_0^{*S} \rightarrow a_{10}^{S} \wedge a_{11}^{S} \wedge a_{12}^{S} \wedge a_{13}^{S} \wedge a_{14}^{S}$$
$$\wedge a_{15}^{S} \wedge a_{16}^{S} \wedge a_{17}^{S} \wedge a_{18}^{S} \wedge a_{19}^{S}$$
$$\wedge a_{20}^{S} \wedge a_{21}^{S} \wedge a_{22}^{S} \wedge a_{23}^{S} \wedge a_{24}^{S} \wedge a_{25}^{S}$$

Sensor Validation and Proactive Fault Analysis Diagnostic Rules

$$P_0^{*H} \rightarrow \left(a_{10}{}^H \vee a_{11}{}^H \vee a_{12}{}^L \vee a_{13}{}^H \vee a_{14}{}^L \vee a_{15}{}^L \vee a_{16}{}^H \vee a_{18}{}^H \vee a_{19}{}^H\right.$$
$$\left. \vee\, a_{21}{}^H \vee a_{22}{}^H \vee a_{25}{}^H\right) \wedge a_{17}{}^S \wedge a_{20}{}^S \wedge a_{23}{}^S \wedge a_{24}{}^S$$

$$P_0^{*L} \rightarrow \left(a_{10}{}^L \vee a_{11}{}^L \vee a_{12}{}^H \vee a_{13}{}^L \vee a_{14}{}^H \vee a_{15}{}^H \vee a_{16}{}^L \vee a_{20}{}^L \vee a_{23}{}^L\right.$$
$$\left. \vee\, a_{24}{}^L \vee a_{25}{}^L\right) \wedge a_{17}{}^S \wedge a_{18}{}^S \wedge a_{19}{}^S \wedge a_{21}{}^S \wedge a_{22}{}^S$$

Key performance indicator (KPI) *equation* PE$_1$ below can be formulated to evaluate the actual overall heat transfer coefficient. Consider the following energy balance:

$$\textbf{PE}_1^*: \qquad 0 = F_{Pi}cp_P(T_{Pi} - T_{Po}) - \frac{U_{act}A[(T_{Pi} - T_{Wo}) - (T_{Po} - T_{Wi})]}{\ln[(T_{Pi} - T_{Wo})/(T_{Po} - T_{Wi})]}$$

where

U_{act} = value of actual overall heat transfer coefficient as currently occurring in the heat exchanger

A = actual heat transfer surface area of a shell-and-tube heat exchanger

All primary models as defined by equation (2.1) are one model with no unknown assumption variables. Since U_{act} does not have a standard value that can always be substituted reliably into PE$_1$ to close it, this situation is one equation in one unknown. Consequently, it thus cannot be effectively used as a primary model by MOME. PE$_1$ can be solved for the overall heat transfer coefficient as follows:

$$\textbf{PE}_1: \quad U_{act} = \frac{F_{Pi}cp_P(T_{Pi} - T_{Po})}{A\{[(T_{Pi} - T_{Wo}) - (T_{Po} - T_{Wi})]/\ln[(T_{Pi} - T_{Wo})/(T_{Po} - T_{Wi})]\}}$$

Before a meaningful calculation of this KPI (i.e., the overall heat transfer coefficient) is possible, the sensor data required to do this calculation must be validated. Otherwise, incorrect decisions about the current level of fouling could occur. This forms a fault precedence hierarchy within the diagnostic logic for monitoring all KPIs. For this example it would be as follows:

$$a_{10}{}^S \wedge a_{11}{}^S \wedge a_{12}{}^S \wedge a_{13}{}^S \wedge a_{14}{}^S \wedge a_{15}{}^S \wedge (U_{act} < U_{fouled})$$
$$\rightarrow \text{heat exchanger is fouled}$$

where

U_{fouled} = sufficiently degraded heat transfer performance for the heat exchanger to be considered fouled

Thus, only after all the sensor measurements used in the calculation of the overall heat transfer coefficient are validated can this calculation be believed to be an accurate metric of current process operations. All KPIs (defined in FALCONEER™ IV as performance equations) have all the sensor measurements that are both used directly in their calculation and are present in the primary models defined in the SV&PFA module validated prior to these KPI values being alarmed. If any of these sensor measurements are suspect, the program automatically alarms those KPIs as calculation-invalid.

Before proceeding further, one more issue concerning how FALCONEER™ IV actually handles lower precedence fault situations (i.e., those calculating performance equations as shown above) needs to be addressed. The original FMC FALCONEER KBS actually contained explicit sensor validation diagnostic rules like those shown in the two examples above; FALCONEER™ IV KBS applications do not. Instead, they rely on the following logically equivalent reasoning to make the identical conclusions.

In the case of the Boolean logic representation presented so far, the following identity[11] always holds:

$$a_i{}^L \ominus a_i{}^S \ominus a_i{}^H$$

This leads directly to the following statement:

$$\left(\neg a_i{}^L \wedge \neg a_i{}^H\right) \leftrightarrow a_i{}^S$$

Consequently, the diagnostic rule above for reasoning about the performance equation for the overall heat transfer coefficient can be written equivalently as follows:

$$\left(\neg a_{10}{}^L \wedge \neg a_{10}{}^H\right) \wedge \left(\neg a_{11}{}^L \wedge \neg a_{11}{}^H\right) \wedge \left(\neg a_{12}{}^L \wedge \neg a_{12}{}^H\right) \wedge \left(\neg a_{13}{}^L \wedge \neg a_{13}{}^H\right)$$
$$\wedge \left(\neg a_{14}{}^L \wedge \neg a_{14}{}^H\right) \wedge \left(\neg a_{15}{}^L \wedge \neg a_{15}{}^H\right) \wedge \left(U_{\text{act}} < U_{\text{fouled}}\right)$$
$$\rightarrow \text{heat exchanger is fouled}$$

---

[11]Similarly, $P_i{}^L \ominus P_i{}^S \ominus P_i{}^H$ and $S_i{}^L \ominus S_i{}^S \ominus S_i{}^H$ also hold. The latter two identities have significance when logically specifying SV&PFA diagnostic rules for particular multiple-fault situations (see Example 3.3).

Thus, the MOME diagnostic strategy as implemented in FALCONEER™ IV validates a given possible assumption variable by default; that is, if the other various SV&PFA diagnostic rules cannot determine that a particular assumption deviation is currently occurring (neither high nor low, if applicable, at the lowest level of possible diagnostic resolution), the logical implication is that the assumption variable is satisfied. This use of default reasoning considerably simplifies the generalized algorithm required to implement the MOME diagnostic strategy to focus on only determining which possible assumption variable deviations (both single and multiple, as demonstrated in the following two examples) are currently occurring.

3.3.3   Example of MOME SV&PFA Diagnostic Rules with Measurement Redundancy

Example 3.3   If redundant measurements of a process variable are available, it is sometimes possible to validate both directly or else directly detect which of the measurements is errant. With the MOME strategy, there is no need to poll (i.e., require three or more redundant measurements of which two or more must agree to indicate the "valid value"). Instead, physical redundancy is incorporated directly with the analytical redundancy techniques described above to derive SV&PFA diagnostic rules. The following example demonstrates how this can be accomplished.

Consider the instrumented pipeline shown in Figure 3.1. The following four primary models, $\mathbf{P_6}$ to $\mathbf{P_9}$ below, can be derived directly with these five modeling assumption variables:

(1) a_{F1A}^{S}:   Linear; flowmeter F_{1A} measurement is correct
(2) a_{F1B}^{S}:   Linear; flowmeter F_{1B} measurement is correct
(3) a_{F2A}^{S}:   Linear; flowmeter F_{2A} measurement is correct
(4) a_{F2B}^{S}:   Linear; flowmeter F_{2B} measurement is correct
(5) a_{NoLeak}^{S}:   Linear; no process leak (L) is occurring in the pipeline

Figure 3.1   Redundantly instrumented process pipeline.

$$\mathbf{P_6}: \qquad 0 = F_{1A} - F_{2A} - L$$
$$\mathbf{P_7}: \qquad 0 = F_{1A} - F_{2B} - L$$
$$\mathbf{P_8}: \qquad 0 = F_{1B} - F_{2A} - L$$
$$\mathbf{P_9}: \qquad 0 = F_{1B} - F_{2B} - L$$

The following two secondary models can be created directly from these primary models.

$$\mathbf{S_4}: \qquad 0 = F_{1A} - F_{1B}$$
$$\mathbf{S_5}: \qquad 0 = F_{2A} - F_{2B}$$

The following sensor validation diagnostic rule can be written:

$$P_6^S \wedge P_7^S \wedge P_8^S \wedge P_9^S \wedge S_4^S \wedge S_5^S \rightarrow a_{F1A}^S \wedge a_{F1B}^S \wedge a_{F2A}^S \wedge a_{F2B}^S \wedge a_{NoLeak}^S$$

The following SV&PFA diagnostic rules can be derived directly for identifying single-fault situations at the highest level of diagnostic resolution[12]:

$$P_6^H \wedge P_7^H \wedge P_8^H \wedge P_9^H \wedge S_4^S \wedge S_5^S \rightarrow a_{F1A}^S \wedge a_{F1B}^S \wedge a_{F2A}^S \wedge a_{F2B}^S \wedge a_{NoLeak}^H$$

$$P_6^H \wedge P_7^H \wedge P_8^S \wedge P_9^S \wedge S_4^H \wedge S_5^S \rightarrow a_{F1A}^H \wedge a_{F1B}^S \wedge a_{F2A}^S \wedge a_{F2B}^S \wedge a_{NoLeak}^S$$

$$P_6^L \wedge P_7^L \wedge P_8^S \wedge P_9^S \wedge S_4^L \wedge S_5^S \rightarrow a_{F1A}^L \wedge a_{F1B}^S \wedge a_{F2A}^S \wedge a_{F2B}^S \wedge a_{NoLeak}^S$$

$$P_6^S \wedge P_7^S \wedge P_8^H \wedge P_9^H \wedge S_4^L \wedge S_5^S \rightarrow a_{F1B}^H \wedge a_{F1A}^S \wedge a_{F2A}^S \wedge a_{F2B}^S \wedge a_{NoLeak}^S$$

$$P_6^S \wedge P_7^S \wedge P_8^L \wedge P_9^L \wedge S_4^H \wedge S_5^S \rightarrow a_{F1B}^L \wedge a_{F1A}^S \wedge a_{F2A}^S \wedge a_{F2B}^S \wedge a_{NoLeak}^S$$

$$P_6^H \wedge P_7^S \wedge P_8^H \wedge P_9^S \wedge S_4^S \wedge S_5^L \rightarrow a_{F2A}^L \wedge a_{F1A}^S \wedge a_{F1B}^S \wedge a_{F2B}^S \wedge a_{NoLeak}^S$$

$$P_6^L \wedge P_7^S \wedge P_8^L \wedge P_9^S \wedge S_4^S \wedge S_5^H \rightarrow a_{F2A}^H \wedge a_{F1A}^S \wedge a_{F1B}^S \wedge a_{F2B}^S \wedge a_{NoLeak}^S$$

$$P_6^S \wedge P_7^H \wedge P_8^S \wedge P_9^H \wedge S_4^S \wedge S_5^H \rightarrow a_{F2B}^L \wedge a_{F1B}^S \wedge a_{F1A}^S \wedge a_{F2A}^S \wedge a_{NoLeak}^S$$

$$P_6^S \wedge P_7^L \wedge P_8^S \wedge P_9^L \wedge S_4^S \wedge S_5^L \rightarrow a_{F2B}^H \wedge a_{F1B}^S \wedge a_{F1A}^S \wedge a_{F2A}^S \wedge a_{NoLeak}^S$$

---

[12]This example assumes that intermediate assumption variable deviations do not occur (i.e., all models have the same value of their given theoretical minimum assumption variable deviations, and thus lower-resolution single-fault SV&PFA diagnostic rules are not necessary). Consequently, the full pattern of model violations occurs whenever there is a significant assumption variable deviation. Typically, this should normally be true for such redundant models. Nonetheless, our process fault analysis program automatically considers all possible cases of intermediate assumption deviations with lower-resolution diagnostic rules.

The following SV&PFA diagnostic rules demonstrate the various interactive multiple-fault situations,[13] which can be identified directly with MOME diagnostic logic:

$$P_6^H \land P_7^H \land P_8^H \land P_9^H \land S_5^S \rightarrow a_{F1A}^H \land a_{F1B}^H \land a_{F2A}^S \land a_{F2B}^S \land a_{NoLeak}^S$$

$$P_6^H \land P_7^H \land P_8^L \land P_9^L \land S_4^H \land S_5^S \rightarrow a_{F1A}^H \land a_{F1B}^L \land a_{F2A}^S \land a_{F2B}^S \land a_{NoLeak}^S$$

$$P_7^H \land P_8^L \land P_9^S \land S_4^H \land S_5^H \rightarrow a_{F1A}^H \land a_{F1B}^S \land a_{F2A}^H \land a_{F2B}^S \land a_{NoLeak}^S$$

$$P_6^H \land P_7^H \land P_8^H \land P_9^S \land S_4^H \land S_5^L \rightarrow a_{F1A}^H \land a_{F1B}^S \land a_{F2A}^L \land a_{F2B}^S \land a_{NoLeak}^S$$

$$P_6^H \land P_8^S \land P_9^L \land S_4^H \land S_5^L \rightarrow a_{F1A}^H \land a_{F1B}^S \land a_{F2A}^S \land a_{F2B}^H \land a_{NoLeak}^S$$

$$P_6^H \land P_7^H \land P_8^S \land P_9^H \land S_4^H \land S_5^H \rightarrow a_{F1A}^H \land a_{F1B}^S \land a_{F2A}^S \land a_{F2B}^L \land a_{NoLeak}^S$$

$$P_6^H \land P_7^H \land P_8^H \land P_9^H \land S_4^H \land S_5^S \rightarrow a_{F1A}^H \land a_{F1B}^S \land a_{F2A}^S \land a_{F2B}^S \land a_{NoLeak}^H$$

$$P_6^L \land P_7^L \land P_8^H \land P_9^H \land S_4^L \land S_5^S \rightarrow a_{F1A}^L \land a_{F1B}^H \land a_{F2A}^S \land a_{F2B}^S \land a_{NoLeak}^S$$

$$P_6^L \land P_7^L \land P_8^L \land P_9^L \land S_5^S \rightarrow a_{F1A}^L \land a_{F1B}^L \land a_{F2A}^S \land a_{F2B}^S \land a_{NoLeak}^S$$

$$P_6^L \land P_7^L \land P_8^L \land P_9^S \land S_4^L \land S_5^H \rightarrow a_{F1A}^L \land a_{F1B}^S \land a_{F2A}^H \land a_{F2B}^S \land a_{NoLeak}^S$$

$$P_7^L \land P_8^H \land P_9^S \land S_4^L \land S_5^L \rightarrow a_{F1A}^L \land a_{F1B}^S \land a_{F2A}^L \land a_{F2B}^S \land a_{NoLeak}^S$$

$$P_6^L \land P_7^L \land P_8^S \land P_9^L \land S_4^L \land S_5^L \rightarrow a_{F1A}^L \land a_{F1B}^S \land a_{F2A}^S \land a_{F2B}^H \land a_{NoLeak}^S$$

$$P_6^L \land P_8^S \land P_9^H \land S_4^L \land S_5^H \rightarrow a_{F1A}^L \land a_{F1B}^S \land a_{F2A}^S \land a_{F2B}^L \land a_{NoLeak}^S$$

$$P_8^H \land P_9^H \land S_4^L \land S_5^S \rightarrow a_{F1A}^L \land a_{F1B}^S \land a_{F2A}^S \land a_{F2B}^S \land a_{NoLeak}^H$$

$$P_6^L \land P_7^S \land P_9^H \land S_4^L \land S_5^H \rightarrow a_{F1A}^S \land a_{F1B}^H \land a_{F2A}^H \land a_{F2B}^S \land a_{NoLeak}^S$$

$$P_6^H \land P_7^S \land P_8^H \land P_9^H \land S_4^L \land S_5^L \rightarrow a_{F1A}^S \land a_{F1B}^H \land a_{F2A}^L \land a_{F2B}^S \land a_{NoLeak}^S$$

$$P_6^S \land P_7^L \land P_8^H \land S_4^L \land S_5^L \rightarrow a_{F1A}^S \land a_{F1B}^H \land a_{F2A}^S \land a_{F2B}^H \land a_{NoLeak}^S$$

$$P_6^S \land P_7^H \land P_8^H \land P_9^H \land S_4^L \land S_5^H \rightarrow a_{F1A}^S \land a_{F1B}^H \land a_{F2A}^S \land a_{F2B}^L \land a_{NoLeak}^S$$

$$P_6^H \land P_7^H \land P_8^H \land P_9^H \land S_4^L \land S_5^S \rightarrow a_{F1A}^S \land a_{F1B}^H \land a_{F2A}^S \land a_{F2B}^S \land a_{NoLeak}^H$$

$$P_6^L \land P_7^S \land P_8^L \land P_9^L \land S_4^H \land S_5^H \rightarrow a_{F1A}^S \land a_{F1B}^L \land a_{F2A}^H \land a_{F2B}^S \land a_{NoLeak}^S$$

$$P_6^H \land P_7^S \land P_9^L \land S_4^H \land S_5^L \rightarrow a_{F1A}^S \land a_{F1B}^L \land a_{F2A}^L \land a_{F2B}^S \land a_{NoLeak}^S$$

$$P_6^S \land P_7^L \land P_8^L \land P_9^L \land S_4^H \land S_5^L \rightarrow a_{F1A}^S \land a_{F1B}^L \land a_{F2A}^S \land a_{F2B}^H \land a_{NoLeak}^S$$

$$P_6^S \land P_7^H \land P_8^L \land S_4^H \land S_5^H \rightarrow a_{F1A}^S \land a_{F1B}^L \land a_{F2A}^S \land a_{F2B}^L \land a_{NoLeak}^S$$

$$P_6^H \land P_7^H \land S_4^H \land S_5^S \rightarrow a_{F1A}^S \land a_{F1B}^L \land a_{F2A}^S \land a_{F2B}^S \land a_{NoLeak}^H$$

---

[13] The formal definitions of both noninteractive and interactive multiple fault situations are given in Chapter 4. Briefly, noninteractive multiple-fault situations are those for which its member fault situations' relevant model sets' intersection is the empty set. Interactive multiple-fault situations are those in which this intersection is not the empty set.

$$P_6^L \wedge P_7^L \wedge P_8^L \wedge P_9^L \wedge S_4^S \rightarrow a_{F1A}^S \wedge a_{F1B}^S \wedge a_{F2A}^H \wedge a_{F2B}^H \wedge a_{NoLeak}^S$$

$$P_6^L \wedge P_7^H \wedge P_8^L \wedge P_9^H \wedge S_4^S \wedge S_5^H \rightarrow a_{F1A}^S \wedge a_{F1B}^S \wedge a_{F2A}^H \wedge a_{F2B}^L \wedge a_{NoLeak}^S$$

$$P_7^H \wedge P_9^H \wedge S_4^S \wedge S_5^H \rightarrow a_{F1A}^S \wedge a_{F1B}^S \wedge a_{F2A}^H \wedge a_{F2B}^S \wedge a_{NoLeak}^H$$

$$P_6^H \wedge P_7^L \wedge P_8^H \wedge P_9^L \wedge S_4^S \wedge S_5^L \rightarrow a_{F1A}^S \wedge a_{F1B}^S \wedge a_{F2A}^L \wedge a_{F2B}^H \wedge a_{NoLeak}^S$$

$$P_6^H \wedge P_7^H \wedge P_8^H \wedge P_9^H \wedge S_4^S \rightarrow a_{F1A}^S \wedge a_{F1B}^S \wedge a_{F2A}^L \wedge a_{F2B}^L \wedge a_{NoLeak}^S$$

$$P_6^H \wedge P_7^H \wedge P_8^H \wedge P_9^H \wedge S_4^S \wedge S_5^L \rightarrow a_{F1A}^S \wedge a_{F1B}^S \wedge a_{F2A}^L \wedge a_{F2B}^S \wedge a_{NoLeak}^H$$

$$P_6^H \wedge P_8^H \wedge S_4^S \wedge S_5^L \rightarrow a_{F1A}^S \wedge a_{F1B}^S \wedge a_{F2A}^S \wedge a_{F2B}^H \wedge a_{NoLeak}^H$$

$$P_6^H \wedge P_7^H \wedge P_8^H \wedge P_9^H \wedge S_4^S \wedge S_5^H \rightarrow a_{F1A}^S \wedge a_{F1B}^S \wedge a_{F2A}^S \wedge a_{F2B}^L \wedge a_{NoLeak}^H$$

As demonstrated, all possible pairs of interactive multiple-fault situations can be diagnosed directly with these additional SV&PFA diagnostic rules. Unfortunately, the diagnostic rule format presented above makes it possible that the multiple-fault situations indicated can also be diagnosed as single-fault situations. Consider the first SV&PFA diagnostic rules in the lists for both single- and multiple-fault situations. The single-fault rule will fire during the multiple-fault situation indicated whenever S_4^S also occurs. This secondary model can be low, satisfied, or high during such multiple-faults, depending on the relative levels of a_{F1A}^H and a_{F1B}^H.[14] It should be emphasized again that both the single- and multiple-fault hypotheses are all equally logically plausible explanations of the current process state. These issues regarding multiple-fault analysis are discussed further in Chapter 4.

3.3.4 Example of MOME SV&PFA Diagnostic Rules for Interactive Multiple-Faults

Example 3.4   As demonstrated in Example 3.3, interactive multiple-fault situations are those whose individual faults can affect the behavior of one or more of the same relevant primary or secondary models used to diagnose those particular individual component fault situations of the multiple-fault situation. This example demonstrates further the limitations on the diagnostic resolution of the diagnoses of such multiple-fault situations.

Consider the instrumented pipeline depicted in Figure 3.2. With the following assumption variables, the following two primary models and one secondary model can be derived.

---

[14]The following equivalence is always a true statement: $(A \wedge B) \leftrightarrow A$ if and only if B is valid, as in this case; $(S_4^L \ominus S_4^S \ominus S_4^H)$. Such validities can thus be left out of the SV&PFA diagnostic rules.

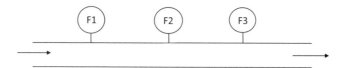

Figure 3.2 Instrumented process pipeline.

(1) $a_{F1}{}^S$:   Linear; flowmeter F_1 measurement is correct
(2) $a_{F2}{}^S$:   Linear; flowmeter F_2 measurement is correct
(3) $a_{F3}{}^S$:   Linear; flowmeter F_3 measurement is correct
(4) $a_{NoLeak1}{}^S$: Linear; no process leak (L_1) is occurring in the first leg of the pipeline
(5) $a_{NoLeak2}{}^S$: Linear; no process leak (L_2) is occurring in the last leg of the pipeline

P_{10}: $0 = F_1 - F_2 - L_1$

P_{11}: $0 = F_2 - F_3 - L_2$

S_6: $0 = F_1 - F_3 - L_1 - L_2$

The following sensor validation diagnostic rule can be derived:

$$P_{10}{}^S \wedge P_{11}{}^S \wedge S_6{}^S \rightarrow a_{F1}{}^S \wedge a_{F2}{}^S \wedge a_{F3}{}^S \wedge a_{NoLeak1}{}^S \wedge a_{NoLeak2}{}^S$$

The following highest-resolution-level SV&PFA diagnostic rules for potential single-fault situations can readily be derived:

$$P_{10}{}^H \wedge P_{11}{}^L \wedge S_6{}^S \rightarrow a_{F1}{}^S \wedge a_{F2}{}^L \wedge a_{F3}{}^S \wedge a_{NoLeak1}{}^S \wedge a_{NoLeak2}{}^S$$

$$P_{10}{}^L \wedge P_{11}{}^H \wedge S_6{}^S \rightarrow a_{F1}{}^S \wedge a_{F2}{}^H \wedge a_{F3}{}^S \wedge a_{NoLeak1}{}^S \wedge a_{NoLeak2}{}^S$$

$$P_{10}{}^H \wedge P_{11}{}^S \wedge S_6{}^H \rightarrow (a_{F1}{}^H \vee a_{NoLeak1}{}^H) \wedge a_{F2}{}^S \wedge a_{F3}{}^S \wedge a_{NoLeak2}{}^S$$

$$P_{10}{}^L \wedge P_{11}{}^S \wedge S_6{}^L \rightarrow a_{F1}{}^L \wedge a_{F2}{}^S \wedge a_{F3}{}^S \wedge a_{NoLeak1}{}^S \wedge a_{NoLeak2}{}^S$$

$$P_{10}{}^S \wedge P_{11}{}^L \wedge S_6{}^L \rightarrow a_{F1}{}^S \wedge a_{F2}{}^S \wedge a_{F3}{}^H \wedge a_{NoLeak1}{}^S \wedge a_{NoLeak2}{}^S$$

$$P_{10}{}^S \wedge P_{11}{}^H \wedge S_6{}^H \rightarrow a_{F1}{}^S \wedge a_{F2}{}^S \wedge (a_{F3}{}^L \vee a_{NoLeak2}{}^H) \wedge a_{NoLeak1}{}^S$$

Similarly, lower-resolution SV&PFA single-fault diagnostic rules can also readily be derived:

$$P_{10}{}^H \wedge P_{11}{}^S \wedge S_6{}^S \rightarrow (a_{F1}{}^H \vee a_{F2}{}^L \vee a_{NoLeak1}{}^H) \wedge a_{F3}{}^S \wedge a_{NoLeak2}{}^S$$

$$P_{10}{}^L \wedge P_{11}{}^S \wedge S_6{}^S \rightarrow (a_{F1}{}^L \vee a_{F2}{}^H) \wedge a_{F3}{}^S \wedge a_{NoLeak1}{}^S \wedge a_{NoLeak2}{}^S$$

$$P_{10}{}^S \wedge P_{11}{}^H \wedge S_6{}^S \rightarrow (a_{F2}{}^H \vee a_{F3}{}^L \vee a_{NoLeak2}{}^H) \wedge a_{F1}{}^S \wedge a_{NoLeak1}{}^S$$

$$P_{10}{}^S \wedge P_{11}{}^L \wedge S_6{}^S \rightarrow (a_{F2}{}^L \vee a_{F3}{}^H) \wedge a_{F1}{}^S \wedge a_{NoLeak1}{}^S \wedge a_{NoLeak2}{}^S$$

The following two lower-resolution SV&PFA diagnostic rules could also be derived but are currently excluded from our analysis. By preference, we require at least one primary model violation to occur in order to make a diagnosis. This makes the resulting diagnoses more meaningful because the primary models are specifically chosen to be the most fundamental representations of normal process behavior, although sometimes the excluded SV&PFA diagnostic rules can be more sensitive to the fault hypotheses indicated.

$$P_{10}{}^S \wedge P_{11}{}^S \wedge S_6{}^H \rightarrow (a_{F1}{}^H \vee a_{F3}{}^L \vee a_{NoLeak1}{}^H \vee a_{NoLeak2}{}^H) \wedge a_{F2}{}^S$$

$$P_{10}{}^S \wedge P_{11}{}^S \wedge S_6{}^L \rightarrow (a_{F1}{}^L \vee a_{F3}{}^H) \wedge a_{F2}{}^S \wedge a_{NoLeak1}{}^S \wedge a_{NoLeak2}{}^S$$

Employing *Occam's razor*,[15] any currently occurring situation in the process system which created patterns of diagnostic evidence matching any of the first 11 patterns above would be diagnosed as indicated. However, as shown in Example 3.3, these 11 SV&PFA diagnostic rules can also possibly diagnose many particular multiple-fault situation manifestations. If not diagnosed explicitly, the fault analyzer would misdiagnose them whenever they occurred. Furthermore, it should be reiterated that because of the inclusive OR logic identifying faults in some of the highest and especially lower-resolution rules, it is always the case that the fault hypothesis is one or more of the faults identified, or possibly all faults identified are currently occurring. All fault hypotheses are considered equally logically plausible explanations of the current process operating state.

Furthermore, all of the other possible patterns of residual behavior [i.e., $27 (=3^3)$ possible patterns -13 patterns shown above $= 14$ other possible patterns] would indicate directly that a multiple-fault situation was taking place. The following SV&PFA diagnostic rules could be derived directly to diagnose those situations:

$$P_{10}{}^S \wedge P_{11}{}^L \wedge S_6{}^H \rightarrow \left((a_{F1}{}^H \wedge a_{F2}{}^L) \wedge a_{F3}{}^S \wedge a_{NoLeak1}{}^S \wedge a_{NoLeak2}{}^S\right) \quad \ominus$$

$$\left((a_{F1}{}^H \wedge a_{F3}{}^H) \wedge a_{F2}{}^S \wedge a_{NoLeak1}{}^S \wedge a_{NoLeak2}{}^S\right) \quad \ominus$$

$$\left((a_{F2}{}^L \wedge a_{F3}{}^L) \wedge a_{F1}{}^S \wedge a_{NoLeak1}{}^S \wedge a_{NoLeak2}{}^S\right) \quad \ominus$$

---

[15]Also referred to as the *principle of parsimony* [9]. This principle is described more fully in Chapter 4. Briefly, it recommends that the simplest possible explanation of the available evidence be the preferred hypothesis.

$$\left(\left(a_{F2}{}^{L} \land a_{NoLeak1}{}^{H}\right) \land a_{F1}{}^{S} \land a_{F3}{}^{S} \land a_{NoLeak2}{}^{S}\right) \quad \ominus$$

$$\left(\left(a_{F2}{}^{L} \land a_{NoLeak2}{}^{H}\right) \land a_{F1}{}^{S} \land a_{F3}{}^{S} \land a_{NoLeak1}{}^{S}\right) \quad \ominus$$

$$\left(\left(a_{F3}{}^{H} \land a_{NoLeak1}{}^{H}\right) \land a_{F1}{}^{S} \land a_{F2}{}^{S} \land a_{NoLeak2}{}^{S}\right)$$

$$P_{10}{}^{S} \land P_{11}{}^{H} \land S_{6}{}^{L} \rightarrow \left(\left(a_{F1}{}^{L} \land a_{F2}{}^{H}\right) \land a_{F3}{}^{S} \land a_{NoLeak1}{}^{S} \land a_{NoLeak2}{}^{S}\right) \quad \ominus$$

$$\left(\left(a_{F1}{}^{L} \land a_{F3}{}^{L}\right) \land a_{F2}{}^{S} \land a_{NoLeak1}{}^{S} \land a_{NoLeak2}{}^{S}\right) \quad \ominus$$

$$\left(\left(a_{F2}{}^{H} \land a_{F3}{}^{H}\right) \land a_{F1}{}^{S} \land a_{NoLeak1}{}^{S} \land a_{NoLeak2}{}^{S}\right) \quad \ominus$$

$$\left(\left(a_{F1}{}^{L} \land a_{NoLeak2}{}^{H}\right) \land a_{F2}{}^{S} \land a_{F3}{}^{S} \land a_{NoLeak1}{}^{S}\right)$$

$$P_{10}{}^{L} \land P_{11}{}^{S} \land S_{6}{}^{H} \rightarrow \left(\left(a_{F1}{}^{H} \land a_{F2}{}^{H}\right) \land a_{F3}{}^{S} \land a_{NoLeak1}{}^{S} \land a_{NoLeak2}{}^{S}\right) \quad \ominus$$

$$\left(\left(a_{F1}{}^{L} \land a_{F3}{}^{L}\right) \land a_{F2}{}^{S} \land a_{NoLeak1}{}^{S} \land a_{NoLeak2}{}^{S}\right) \quad \ominus$$

$$\left(\left(a_{F2}{}^{H} \land a_{F3}{}^{L}\right) \land a_{F1}{}^{S} \land a_{NoLeak1}{}^{S} \land a_{NoLeak2}{}^{S}\right) \quad \ominus$$

$$\left(\left(a_{F1}{}^{L} \land a_{NoLeak2}{}^{H}\right) \land a_{F2}{}^{S} \land a_{F3}{}^{S} \land a_{NoLeak1}{}^{S}\right) \quad \ominus$$

$$\left(\left(a_{F2}{}^{H} \land a_{NoLeak1}{}^{H}\right) \land a_{F1}{}^{S} \land a_{F3}{}^{S} \land a_{NoLeak2}{}^{S}\right) \quad \ominus$$

$$\left(\left(a_{F2}{}^{H} \land a_{NoLeak2}{}^{H}\right) \land a_{F1}{}^{S} \land a_{F3}{}^{S} \land a_{NoLeak1}{}^{S}\right)$$

$$P_{10}{}^{L} \land P_{11}{}^{L} \land S_{6}{}^{S} \rightarrow \left(\left(a_{F1}{}^{L} \land a_{F2}{}^{L}\right) \land a_{F3}{}^{S} \land a_{NoLeak1}{}^{S} \land a_{NoLeak2}{}^{S}\right) \quad \ominus$$

$$\left(\left(a_{F1}{}^{L} \land a_{F3}{}^{H}\right) \land a_{F2}{}^{S} \land a_{NoLeak1}{}^{S} \land a_{NoLeak2}{}^{S}\right) \quad \ominus$$

$$\left(\left(a_{F2}{}^{H} \land a_{F3}{}^{H}\right) \land a_{F1}{}^{S} \land a_{NoLeak1}{}^{S} \land a_{NoLeak2}{}^{S}\right)$$

$$P_{10}{}^{L} \land P_{11}{}^{L} \land S_{6}{}^{L} \rightarrow \left(\left(a_{F1}{}^{L} \land a_{F2}{}^{L}\right) \land a_{F3}{}^{S} \land a_{NoLeak1}{}^{S} \land a_{NoLeak2}{}^{S}\right) \quad \ominus$$

$$\left(\left(a_{F1}{}^{L} \land a_{F3}{}^{H}\right) \land a_{F2}{}^{S} \land a_{NoLeak1}{}^{S} \land a_{NoLeak2}{}^{S}\right) \quad \ominus$$

$$\left(\left(a_{F2}{}^{H} \land a_{F3}{}^{H}\right) \land a_{F1}{}^{S} \land a_{NoLeak1}{}^{S} \land a_{NoLeak2}{}^{S}\right)$$

$$P_{10}{}^{L} \land P_{11}{}^{H} \land S_{6}{}^{L} \rightarrow \left(\left(a_{F1}{}^{L} \land a_{F2}{}^{H}\right) \land a_{F3}{}^{S} \land a_{NoLeak1}{}^{S} \land a_{NoLeak2}{}^{S}\right) \quad \ominus$$

$$\left(\left(a_{F1}{}^{L} \land a_{F3}{}^{L}\right) \land a_{F2}{}^{S} \land a_{NoLeak1}{}^{S} \land a_{NoLeak2}{}^{S}\right) \quad \ominus$$

$$\left(\left(a_{F2}{}^{H} \land a_{F3}{}^{H}\right) \land a_{F1}{}^{S} \land a_{NoLeak1}{}^{S} \land a_{NoLeak2}{}^{S}\right) \quad \ominus$$

$$\left(\left(a_{F1}{}^{L} \land a_{NoLeak2}{}^{H}\right) \land a_{F2}{}^{S} \land a_{F3}{}^{S} \land a_{NoLeak1}{}^{S}\right)$$

$$P_{10}{}^{L} \land P_{11}{}^{H} \land S_{6}{}^{H} \rightarrow \left(\left(a_{F1}{}^{H} \land a_{F2}{}^{H}\right) \land a_{F3}{}^{S} \land a_{NoLeak1}{}^{S} \land a_{NoLeak2}{}^{S}\right) \quad \ominus$$

$$\left(\left(a_{F1}{}^{L} \land a_{F3}{}^{L}\right) \land a_{F2}{}^{S} \land a_{NoLeak1}{}^{S} \land a_{NoLeak2}{}^{S}\right) \quad \ominus$$

$$((a_{F2}^H \wedge a_{F3}^L) \wedge a_{F1}^S \wedge a_{NoLeak1}^S \wedge a_{NoLeak2}^S) \quad \Theta$$

$$((a_{F1}^L \wedge a_{NoLeak2}^H) \wedge a_{F2}^S \wedge a_{F3}^S \wedge a_{NoLeak1}^S) \quad \Theta$$

$$((a_{F2}^H \wedge a_{NoLeak1}^H) \wedge a_{F1}^S \wedge a_{F3}^S \wedge a_{NoLeak2}^S) \quad \Theta$$

$$((a_{F2}^H \wedge a_{NoLeak2}^H) \wedge a_{F1}^S \wedge a_{F3}^S \wedge a_{NoLeak1}^S)$$

$$P_{10}^H \wedge P_{11}^S \wedge S_6^L \rightarrow ((a_{F1}^L \wedge a_{F2}^L) \wedge a_{F3}^S \wedge a_{NoLeak1}^S \wedge a_{NoLeak2}^S) \quad \Theta$$

$$((a_{F1}^H \wedge a_{F3}^H) \wedge a_{F2}^S \wedge a_{NoLeak1}^S \wedge a_{NoLeak2}^S) \quad \Theta$$

$$((a_{F2}^L \wedge a_{F3}^H) \wedge a_{F1}^S \wedge a_{NoLeak1}^S \wedge a_{NoLeak2}^S) \quad \Theta$$

$$((a_{F3}^H \wedge a_{NoLeak1}^H) \wedge a_{F1}^S \wedge a_{F2}^S \wedge a_{NoLeak2}^S)$$

$$P_{10}^H \wedge P_{11}^L \wedge S_6^L \rightarrow ((a_{F1}^L \wedge a_{F2}^L) \wedge a_{F3}^S \wedge a_{NoLeak1}^S \wedge a_{NoLeak2}^S) \quad \Theta$$

$$((a_{F1}^H \wedge a_{F3}^H) \wedge a_{F2}^S \wedge a_{NoLeak1}^S \wedge a_{NoLeak2}^S) \quad \Theta$$

$$((a_{F2}^L \wedge a_{F3}^H) \wedge a_{F1}^S \wedge a_{NoLeak1}^S \wedge a_{NoLeak2}^S) \quad \Theta$$

$$((a_{F3}^H \wedge a_{NoLeak1}^H) \wedge a_{F1}^S \wedge a_{F2}^S \wedge a_{NoLeak2}^S)$$

$$P_{10}^H \wedge P_{11}^L \wedge S_6^H \rightarrow ((a_{F1}^H \wedge a_{F2}^L) \wedge a_{F3}^S \wedge a_{NoLeak1}^S \wedge a_{NoLeak2}^S) \quad \Theta$$

$$((a_{F1}^H \wedge a_{F3}^H) \wedge a_{F2}^S \wedge a_{NoLeak1}^S \wedge a_{NoLeak2}^S) \quad \Theta$$

$$((a_{F2}^L \wedge a_{F3}^L) \wedge a_{F1}^S \wedge a_{NoLeak1}^S \wedge a_{NoLeak2}^S) \quad \Theta$$

$$((a_{F2}^L \wedge a_{NoLeak2}^H) \wedge a_{F1}^S \wedge a_{F3}^S \wedge a_{NoLeak1}^S) \quad \Theta$$

$$((a_{F2}^L \wedge a_{NoLeak1}^H) \wedge a_{F1}^S \wedge a_{F3}^S \wedge a_{NoLeak2}^S) \quad \Theta$$

$$((a_{F3}^H \wedge a_{NoLeak1}^H) \wedge a_{F1}^S \wedge a_{F2}^S \wedge a_{NoLeak2}^S)$$

$$P_{10}^H \wedge P_{11}^H \wedge S_6^S \rightarrow ((a_{F1}^H \wedge a_{F2}^H) \wedge a_{F3}^S \wedge a_{NoLeak1}^S \wedge a_{NoLeak2}^S) \quad \Theta$$

$$((a_{F1}^H \wedge a_{F3}^L) \wedge a_{F2}^S \wedge a_{NoLeak1}^S \wedge a_{NoLeak2}^S) \quad \Theta$$

$$((a_{F2}^L \wedge a_{F3}^L) \wedge a_{F1}^S \wedge a_{NoLeak1}^S \wedge a_{NoLeak2}^S) \quad \Theta$$

$$((a_{F1}^H \wedge a_{NoLeak2}^H) \wedge a_{F2}^S \wedge a_{F3}^S \wedge a_{NoLeak1}^S) \quad \Theta$$

$$((a_{F2}^L \wedge a_{NoLeak2}^H) \wedge a_{F1}^S \wedge a_{F3}^S \wedge a_{NoLeak1}^S) \quad \Theta$$

$$((a_{F2}^H \wedge a_{NoLeak1}^H) \wedge a_{F1}^S \wedge a_{F3}^S \wedge a_{NoLeak2}^S) \quad \Theta$$

$$((a_{F3}^L \wedge a_{NoLeak1}^H) \wedge a_{F1}^S \wedge a_{F2}^S \wedge a_{NoLeak2}^S) \quad \Theta$$

$$((a_{NoLeak1}^H \wedge a_{NoLeak2}^H) \wedge a_{F1}^S \wedge a_{F2}^S \wedge a_{F3}^S)$$

$$P_{10}{}^H \wedge P_{11}{}^H \wedge S_6{}^H \rightarrow \left(\left(a_{F1}{}^H \wedge a_{F2}{}^H\right) \wedge a_{F3}{}^S \wedge a_{NoLeak1}{}^S \wedge a_{NoLeak2}{}^S\right) \quad \Theta$$

$$\left(\left(a_{F1}{}^H \wedge a_{F3}{}^L\right) \wedge a_{F2}{}^S \wedge a_{NoLeak1}{}^S \wedge a_{NoLeak2}{}^S\right) \quad \Theta$$

$$\left(\left(a_{F2}{}^L \wedge a_{F3}{}^L\right) \wedge a_{F1}{}^S \wedge a_{NoLeak1}{}^S \wedge a_{NoLeak2}{}^S\right) \quad \Theta$$

$$\left(\left(a_{F1}{}^H \wedge a_{NoLeak2}{}^H\right) \wedge a_{F2}{}^S \wedge a_{F3}{}^S \wedge a_{NoLeak1}{}^S\right) \quad \Theta$$

$$\left(\left(a_{F2}{}^L \wedge a_{NoLeak2}{}^H\right) \wedge a_{F1}{}^S \wedge a_{F3}{}^S \wedge a_{NoLeak1}{}^S\right) \quad \Theta$$

$$\left(\left(a_{F2}{}^H \wedge a_{NoLeak1}{}^H\right) \wedge a_{F1}{}^S \wedge a_{F3}{}^S \wedge a_{NoLeak2}{}^S\right) \quad \Theta$$

$$\left(\left(a_{F3}{}^L \wedge a_{NoLeak1}{}^H\right) \wedge a_{F1}{}^S \wedge a_{F2}{}^S \wedge a_{NoLeak2}{}^S\right) \quad \Theta$$

$$\left(\left(a_{NoLeak1}{}^H \wedge a_{NoLeak2}{}^H\right) \wedge a_{F1}{}^S \wedge a_{F2}{}^S \wedge a_{F3}{}^S\right)$$

If either of the following two patterns of diagnostic evidence occurs, at least three simultaneous fault situations must be occurring:

$$P_{10}{}^H \wedge P_{11}{}^H \wedge S_6{}^L \rightarrow \text{three simultaneous faults}$$

$$P_{10}{}^L \wedge P_{11}{}^L \wedge S_6{}^H \rightarrow \text{three simultaneous faults}$$

As is evident from this example, diagnostic resolution in most interactive multiple-fault situations is normally extremely poor; there are just too many combinations of equally plausible explanations that need to be considered. However, it is typically better to list exhaustively all plausible explanations of the current process state than it is to omit less probable scenarios. What is certain from these last 14 SV&PFA diagnostic rules is that whenever they fire, a multiple-fault situation is currently occurring. This topic is discussed in greater detail in Chapter 4.

3.4 GENERAL PROCEDURE FOR DEVELOPING AND VERIFYING COMPETENT MODEL-BASED PROCESS FAULT ANALYZERS

A general procedure for developing and verifying competent model-based process fault analyzers can be summarized as follows:

Step 1. Specify the target process system to be analyzed for process fault situations, along with the fault analyzer's intended scope and its intended operational domain.

Step 2. Derive a set of all possible linearly independent models that accurately describe the behavior of the target process system under malfunction-free (i.e., normal) process operating conditions and calculate statistics (normal offset and variance) for each model as a function of commonly expected process operating conditions (i.e., production rate, etc.).

Step 3. Verify that these models accurately describe the normal process behavior and that the appropriate models become violated when the various possible process operating events contained within the intended scope of the fault analyzer and its intended operational domain occur.

Step 4. Derive all patterns of diagnostic evidence (i.e., the SV&PFA diagnostic rules) anticipated from the behavior expected from these models, which are capable of correctly classifying each possible process operating event contained within the fault analyzer's intended scope.

Step 5. Verify that this set of diagnostic rules (i.e., the diagnostic knowledge base) can competently classify all these events, especially the fault analyzer's target fault situations.

Step 6. Employ an inference procedure efficient enough to search the fault analyzer's knowledge base correctly in real time[16]; this, in turn, allows the fault analysis to be performed continuously online.

As noted, this six-step procedure presents a general outline for creating model-based process fault analyzers. Determining exactly how each of these six steps should be performed depends on the particular model-based diagnostic strategy being employed. So far, we have concentrated on how the various steps should be performed when employing MOME. *Steps 1, 2, and 3 are the only development efforts now currently required to create competent fault analyzers.* Steps 5 is no longer required because step 4 has been automated, allowing step 6 to be a simple exhaustive calculation of all possible process operating events' certainty factors and a simple sort of those results. An algorithm based on a fuzzy logic implementation of MOME is described in detail in Chapter 4 and fully in our patent [12].

3.5 MOME SV&PFA DIAGNOSTIC RULES' LOGIC COMPILER MOTIVATIONS

A critical step in the development of a fault analyzer is verifying that the underlying diagnostic knowledge base performs competently (step 5 of the procedure). Such verification ensures that the SV&PFA diagnostic rules will

---

[16]Real-time performance is said to occur if the program can finish its analysis and report its findings in the time between consecutive process data samplings.

always perform correctly. During development of the FALCON system, this was accomplished by nearly exhaustively testing the fault analyzer with both actual and simulated fault situations. Doing this thoroughly turned out to be the most computationally intensive undertaking of the FALCON project, with more than 5500 hours of actual plant data and about 260 simulated process fault situations being analyzed before the FALCON system was installed online in the plant. The various phases of this verification effort are described in Appendix B and required a period of approximately three years to complete.

Obviously, if such extensive verification efforts were a prerequisite for developing competent fault analyzers in general, these programs would never be used widely in the processing industries. Fortunately, though, efforts similar to that expended in verifying the original FALCON System are not required. It is possible to reduce dramatically the effort required to verify a process fault analyzer based on MOME. If all of the primary process models contained within the declarative knowledge base are well-formulated[17], any subsequent misdiagnoses made by the fault analyzer indicates directly that one or more of the diagnostic rules is incorrect. However, since MOME is a very systematic and logically structured procedure for creating diagnostic rules, it has been possible to codify a fuzzy logic version of it in a compiler program, FALCONEER™ IV, that automatically creates diagnostic knowledge bases based on it. This program ensures that the method of minimal evidence is always applied correctly.

Compilation refers to any process that transforms a representation of knowledge into another representation that can be used more efficiently. Transformation can include optimization as well as the tailoring of representations for potential instruction sets [13]. Compiling is thus a process of knowledge chunking; that is, meaningful portions of knowledge are stored and retrieved as functional units [14]. A compiler is said to create complete solutions if it is capable of producing every possible solution; it is said to be nonredundant if in the course of compiling it produces each solution only once [15]. With the compiler, the most important criterion when choosing the knowledge representation scheme for automating process fault analysis becomes picking one that allows all the necessary knowledge to be represented directly and facilitates its use in compiling the solution of the underlying problem [16]. Since the knowledge underlying automated process fault analysis (i.e., primary models and performance equations) is firm, fixed, and formulated, an algorithmic computer program is more appropriate than a heuristic program [17]. Furthermore, since the problem of automated process fault analysis is characterized by a small search space, mostly reliable data and

---

[17]The concept of well-formulated process models is defined in Chapter 2.

highly reliable knowledge (i.e., evaluated primary models and performance equations), it permits correspondingly simple system architectures; that i.e., systems of this sort can effectively employ exhaustive search [18].

With FALCONEER™ IV, if all the primary process models are indeed well-formulated, the resulting fault analyzer is guaranteed to perform competently. Consequently, verifying the correctness of various diagnostic rules is not necessary: only the correctness of the set of primary process models and performance equations used to create those rules needs to be verified. This in effect converts the much more difficult task of performing competent automated fault analysis into the easier problem of process modeling. The reduction in development effort is substantial.

In the original FALCON system, this would have meant the difference between verifying over 10,000 lines of Common Lisp code (representing the 800+ highly interdependent diagnostic rules and 50+ primary and secondary models and performance equations) and verifying about 100 lines of FORTRAN code (representing just the 20 + primary models and performance equations).[18] The original FALCON KBS could diagnose only single-fault situations at the highest level of diagnostic resolution and only all noninteractive multiple-fault situations.

In the case of the original FALCONEER KBS for the FMC ESP process [19, 20], only 30 primary, 70 secondary models, and 5 performance equations had to be verified rather than the additional hand-compiled 15,000+ SV&PFA diagnostic rules (over 16,000 lines of Visual Basic code). In other words, a comprehensive validation of the SV&PFA diagnostic rules was deemed unnecessary and thus not conducted. This allowed the FALCONEER KBS to be delivered with 1 person-year of effort rather than the approximately 15 person-years required for the original FALCON system development effort. The original FALCONEER KBS could diagnose all single-fault situations at all levels of diagnostic resolution and only all noninteractive multiple-fault situations.

Once FALCONEER™ IV became available [21], converting the original FALCONEER KBS required only coding the 30 primary models and 5 performance equations describing the FMC ESP process. Creating and analyzing the 30+ primary models and 5 performance equations for the FMC LAP process required approximately 2 person-weeks of effort to derive a fully functional and validated process fault analyzer. The benefits to date derived from these two FMC applications have been reported independently by FMC [22]. All FALCONEER™ IV applications can diagnose all single-fault situations at all possible levels of diagnostic resolution and all noninteractive and also almost all possible pairs of interactive multiple-fault situations.

---

[18]Secondary models are created automatically by FALCONEER™ IV.

Consequently, process engineers now only have to derive and maintain the declarative knowledge base containing the various primary process models and performance equations of normal operation. This is advantageous for the following reason. As opposed to other knowledge representation schemes (such as production rules and frames), process models are represented as mathematical equations, an engineer's dream. Furthermore, since all the primary process models within the diagnostic knowledge base are independent of each other, each can be independently added, modified, or removed, as required. The modified database is then merely recompiled to create the improved fault analyzer. This simplifies the maintenance of the process fault analyzer immensely. With the MOME diagnostic logic compiler, the fault analyzer can be improved incrementally with minimal effort and cost as the process system's operating behavior becomes better understood or its topology changes due to process modifications. This greatly improves its maintainability as process operations evolve.

Having FALCONEER™ IV also greatly increases the resulting knowledge-based systems' transparency[19] by separating the domain-specific knowledge (i.e., the various primary models and performance equations of normal operation, constituting the entirety of its declarative knowledge) from its general problem-solving strategy (i.e., the MOME algorithm and exhaustive search of the resulting SV&PFA diagnostic rules). Again, this directly simplifies the much more complicated problem of performing competent automated process fault analysis into an effort of correctly modeling normal process behavior. FALCONEER™ IV thus allows anyone capable of doing such modeling the means of directly creating and maintaining competent and affordable automated process fault analyzers.

3.6 MOME DIAGNOSTIC STRATEGY SUMMARY

The method of minimal evidence (MOME) employs model-based reasoning to logically deduce the cause or causes of abnormal process behavior. It does so with the least amount of diagnostic evidence necessary to uniquely diagnose the various possible fault situations. Moreover, the resulting fault analyzer always makes competent diagnoses at the best resolution and highest sensitivity possible for the given magnitude of the fault(s) occurring.

A key feature of this method is in the way in which the various patterns of diagnostic evidence are selected. This selection fully utilizes all of the information contained within the available diagnostic evidence, especially the estimates of the fault magnitudes of linear assumption variable deviations

---

[19]Transparency is defined as the KBSs' understandability to both that system's developer and the targeted user [23].

inherent in the violated primary model residuals. The strategy followed in this selection relies on default reasoning: all but one of the fault hypotheses (if perfect resolution is possible) being supported by some of the diagnostic evidence contained within a given pattern of diagnostic evidence are shown systematically to be implausible by some of the other evidence also contained within that pattern. Thus, by default, the remaining fault hypothesis is the only plausible explanation of the full pattern of relevant diagnostic evidence. Using default reasoning in this manner allows this diagnostic strategy to base each fault diagnosis on the least amount of diagnostic evidence necessary for that proper diagnosis. This allows many potential multiple-fault situations to be properly diagnosed. This method has been converted into a patented fuzzy logic reasoning algorithm [12] and is now fully automated in the FALCONEER^{TM} IV program [21]. The details of this fuzzy logic [24] algorithm are given in Chapter 4.

REFERENCES

1. Fickelscherer, R. J., *Automated Process Fault Analysis*, Ph.D. dissertation, University of Delaware, Newark, DE, 1990.

2. Fickelscherer, R. J., "A Generalized Approach to Model-Based Process Fault Analysis," in *Proceedings of the 2nd International Conference on Foundations of Computer-Aided Process Operations*, ed. by D. W. T. Rippin, J. C. Hale, and J. F. Davis, CACHE, Inc., Austin, TX, 1994, pp. 451–456.

3. Kramer, M. A., and R. S. H. Mah, "Model-Based Monitoring," in *Foundations of Computer-Aided Process Operations II*, ed. by D. W. T. Rippin, J. C. Hale, and J. F. Davis, CACHE, Inc., Austin, TX, 1994, pp. 45–68.

4. Venkatasubramanian, V., R. Rengaswamy, K. Yin, and S. N. Kavuri, "A Review of Process Fault Detection and Diagnosis: Part 1. Quantitative Model Based Methods," *Computers and Chemical Engineering*, Vol. 27, 2003, pp. 293–311.

5. Charniak, E., and D. McDermott, *Introduction to Artificial Intelligence*, Addison-Wesley, Reading, MA, 1985, pp. 453–455.

6. Parsaye, K., and M. Chignell, *Expert Systems for Experts*, Wiley, New York, 1988, p. 73.

7. Schoning, U., *Logic for Computer Scientists*, 2nd ed., Birkhauser, Boston, 1994, pp. 151–152.

8. Genesereth, M. R., and N. J. Nilsson, *Logical Foundations of Artificial Intelligence*, Morgan Kaufmann Publishers, Los Altos, CA, 1987, pp. 117–121.

9. Reggia, J. A., D. S. Nau, and P. Y. Wang, "Diagnostic Expert Systems Based on a Set Covering Model," in *Developments in Expert Systems*, ed. by M. J. Combs, Academic Press, London, 1984, pp. 35–58.

10. Parsaye, K., and M. Chignell, *Expert Systems for Experts*, Wiley, New York, 1988, pp. 116–117.

11. Davis, R., "Meta-Rules: Reasoning About Control," *Artificial Intelligence*, Vol. 15, 1980, pp. 179–222.

12. Chester, D. L., L. Daniels, R. J. Fickelscherer, and D. H. Lenz, U. S. Patent 7,451,003, "Method and System of Monitoring, Sensor Validation and Predictive Fault Analysis," 2008.

13. Stefik, M., J. Aikins, R. Balzer, J. Benoit, L. Birnbaum, F. and E. Sacerdoti, "The Architecture of Expert Systems," in *Building Expert Systems* ed. by F. Hayes-Roth, D. A. Waterman, and D. B. Lenat, Addison-Wesley, Reading, MA, 1983, p. 121.

14. Harmon, P., and D. King, *Expert Systems: Artificial Intelligence in Business*, Wiley, New York, 1985, p. 30.

15. Stefik, M., J. Aikins, R. Balzer, J. Benoit, L. Birnbaum, F. Hayes-Roth, and E. Sacerdoti, "Basic Concepts for Building Expert Systems," in *Building Expert Systems*, ed. by F. Hayes-Roth, D. A. Waterman, and D. B. Lenat, Addison-Wesley, Reading, MA, 1983, p. 71.

16. Rich, E., *Artificial Intelligence*, McGraw-Hill, New York, 1983, p. 242.

17. Buchanan, B. G., D. Barstow, R. Bechtal, J. Bennet, W. Clancey, C. Kulikowski, T. Mitchell, and D. A. Waterman, "Constructing an Expert System," in *Building Expert Systems*, ed. by F. Hayes-Roth, D. A. Waterman, and D. B. Lenat, Addison-Wesley, Reading, MA, 1983, p. 127.

18. Hayes-Roth, F., D. A. Waterman, and D. B. Lenat, "An Overview of Expert Systems," in *Building Expert Systems*, ed. by F. Hayes-Roth, D. A. Waterman, and D. B. Lenat, Addison-Wesley, Reading, MA, 1983, p. 20.

19. Skotte, R., D. Lenz, R. Fickelscherer, W. An, D. LaphamIII, C. Lymburner, J. Kaylor, D. Baptiste, M. Pinsky, F. Gani, and S. B. Jørgensen, "Advanced Processs Control with Innovation for an Integrated Electrochemical Process," presented at the *AIChE Spring National Meeting*, Houston, TX, 2001.

20. Fickelscherer, R. J., D. H. Lenz, and D. L. Chester, "Intelligent Process Supervision via Automated Data Validation and Fault Analysis: Results of Actual CPI Applications," Paper 115d, presented at the *AIChE Spring National Meeting*, New Orleans, LA, 2003.

21. Fickelscherer, R. J., D. H. Lenz, and D. L. Chester, "Fuzzy Logic Clarifies Operations," *InTech*, October 2005, pp. 53–57.

22. Lymburner, C., J. Rovison, and W. An, "Battling Information Overload," *Control*, September 2006, pp. 95–99.

23. Stefik, M., J. Aikins, R. Balzer, J. Benoit, L. Birnbaum, F. Hayes-Roth, and E. Sacerdoti, "The Architecture of Expert Systems," in *Building Expert Systems*, ed. by F. Hayes-Roth, D. A. Waterman, and D. B. Lenat, Addison-Wesley, Reading, MA, 1983, p. 122.

24. Zadeh, L. A., "Fuzzy Logic," *Computer*, Vol. 21, No. 4, 1988, pp. 83–93.

4

METHOD OF MINIMAL EVIDENCE: FUZZY LOGIC ALGORITHM

4.1 OVERVIEW

In this chapter we describe the details of the fuzzy logic algorithm currently implemented in the FALCONEER™ IV program for performing optimal automated process fault analysis. This implementation generalizes the underlying Boolean logic version of the method of minimal evidence (MOME) described in Chapter 3 to a highly comprehensive fuzzy logic version. This generalization allows for a more compact treatment of potential single- and multiple-fault situations, at all levels of possible diagnostic resolution, with both elegant and efficient uniform sensor validation and proactive fault analysis (SV&PFA) diagnostic rules for diagnosing those situations. This fuzzy logic version of the MOME algorithm thus automates the diagnostic reasoning necessary to perform optimal process fault analysis continuously so that only the underlying well-formulated primary models are required to achieve such performance. Using FALCONEER™ IV consequently simplifies the solution of the more complicated problem of automated process fault analysis into the much simpler problem of adequate process modeling, as described below.

First, all possible primary models (i.e., the set of all possible linearly independent models) describing normal process operating behavior are derived and their associated statistical parameters computed with sufficient actual

Optimal Automated Process Fault Analysis, First Edition.
Richard J. Fickelscherer and Daniel L. Chester.
© 2013 John Wiley & Sons, Inc. Published 2013 by John Wiley & Sons, Inc.

normal process data. This directly defines all the various modeling assumption variables associated with those models and classifies them as either linear or nonlinear assumption variables with regard to those models. The program then serially eliminates all linear assumption variables common to pairs of primary models to derive all of those possible secondary models (i.e., the set of all the linearly dependent models). Next, it determines the patterns of the relevant primary and secondary models' satisfactions and violations expected to occur when the associated modeling assumption variable deviation occurs during process operation according to FALCONEER™ IV's proprietary and patented fuzzy logic MOME diagnostic rules [1, 2]. This is accomplished by examining the first- and second-order partial derivatives of those primary and secondary models with their associated assumption variables. These diagnostic rules are then compared to the actual patterns of model behavior currently occurring in the process to determine the most plausible underlying process operating event(s). Any resulting *intelligent alerts* are immediately brought to the attention of the process operators.

4.2 INTRODUCTION

The original FALCON system derived its conclusions exclusively via the use of Boolean logic. As described in Chapter 2, residuals were determined to be high, low, or satisfied by comparing them against predetermined high and low thresholds. Consequently, a residual could only be in a high, low, or satisfied state at any given instant in time. The diagnostic rule formats described in Chapter 3 optimize the resulting fault analyzer's diagnostic knowledge base for fault analyzers based on Boolean logic. However, Boolean logic places a practical limitation upon the fault analyzer's competency. With Boolean logic, an incremental change in plant state can sometimes radically alter the fault analyzer's conclusions. This occurs if that incremental change in plant state causes the values of one or more of the models' residuals to deviate enough so as to change their Boolean assignment. Kramer [3] refers to this problem as *diagnostic instability*. He argues that it is an inherent problem in the performance of any fault analyzer whose reasoning is based on Boolean logic, regardless of the diagnostic strategy employed to create the diagnostic rules.

Kramer [3] suggests that the best way to eliminate diagnostic instability is to base the conclusions of the process fault analysis on non-Boolean logic. In other knowledge-based system applications, many non-Boolean reasoning methods have been proposed for measuring the degree of fit between the prototypical patterns composing the antecedent of the various production rules and the data being analyzed. One of the most popular of these non-Boolean methods is the calculation of certainty factors. Certainty factors are typically continuous functions that act as empirical measurements of the certainty associated with the degree of fit that exists between a given production rule's antecedent and the data being analyzed. An additional certainty factor can also be associated with the conclusion of a production rule. In this capacity, the additional certainty factor represents the highest certainty that can be placed upon the conclusion of that production rule given that a perfect match exists between its antecedent and the data being analyzed.[1] Many different strategies have been proposed for the combination and manipulation of certainty factors [4–13].

The relative values of the calculated certainty factors act as a metric for comparing the various possible conclusions of the production rules. Typically, the absolute values of these certainty factors do not have any direct theoretical significance [7–11], and for ease of comparison are usually normalized. Fortunately, the details of the certainty factor calculations associated with the diagnostic rules matter less than the semantic and structural content of those rules themselves [14].

The values of certainty factors associated with a given conclusion usually range from −1.0 to 1.0 or 0.0 to 1.0. The higher value in these pairs (i.e., 1.0) indicates that the conclusion is completely certain. Conversely, the lower value indicates either (1) that the conclusion of the production rule is not supported by the evidence present in the data (i.e., 0.0), or (2) that the negation of the production rule's conclusion is certain (i.e., −1.0).[2] Values of the certainty factors between these extreme values tend to indicate that the degree of support derived from the data is somewhere intermediate between the two possible extreme conclusions, with the actual value of the certainty factor acting as a relative measure of which conclusion is more strongly supported.

In model-based process fault analysis, two separate issues need to be addressed concerning the use of certainty factors as a reasoning mechanism. The first concerns the conversion of a model residual into a metric of whether its present value is considered low, normal, or high. Again, this conversion can also be accomplished in a variety of ways [3,10,12,13].[3] The second issue regarding the use of certainty factors as a reasoning mechanism is how to combine the certainty factors associated with the various model violations

---

[1]There is no unique or agreed upon method for dealing with uncertainties in knowledge-based systems [4].

[2]A certainty value of 0.0 when the range is −1.0 to 1.0 typically represents the unknown state. In the following discussion, we will be concerned only with certainty values ranging between 0.0 and 1.0.

[3]The following analysis will be concerned only with the conversion of process model residuals that have normal distributions, such as those that typically result from the evaluation of a model based on one of the conservation laws.

and satisfactions to derive a certainty factor associated with a given fault hypothesis. Both of these issues are addressed directly by the certainty factor calculation described below.

4.3 FUZZY LOGIC OVERVIEW

Rather than using Boolean logic, our model-based diagnostic strategy uses the fuzzy logic [13] certainty factor calculation algorithm described below to make its decisions. This algorithm directly eliminates "diagnostic instability" and minimizes the possible ambiguity in the SV&PFA diagnostic rules. Since the resulting certainty factors are continuous functions, diagnostic instability does not occur. Furthermore, each possible fault hypothesis is diagnosed by considering all possible diagnostic evidence for it in just one diagnostic rule containing all possible patterns of relevant primary and secondary model residual states (i.e., only those that are pertinent to that particular fault situation). Credence in the possible associated low, high, or satisfied states of the various modeling assumption variables required are each assigned values from any and all relevant primary and secondary residual values. This directly optimizes the diagnostic sensitivity along with diagnostic resolution as described for the various SV&PFA diagnostic rules shown in Chapter 3.

Performing automated process fault analysis according to the MOME diagnostic strategy is in effect solving what Newell defines as a well-structured problem [15]; that is, it can be stated in terms of numerical variables (evaluated and interpreted primary and secondary process models), its goals can be specified in terms of a well-defined objective function (classifying a subset of potential process faults) and there exists an algorithmic solution (MOME). Furthermore, since we strive immediately to recognize the implication of all current evidence and all plausible solutions are required, it is recommended that an exhaustive forward chaining search be employed [16].

As implemented in our fuzzy logic algorithm, each diagnostic rule encapsulates a chunk of domain-specific information identifying a specific fault situation if the conditions specified in its premise are fulfilled. The rules are judgmental; that is, they make inexact inferences. As such, every rule premise is a conjunction of clauses, containing arbitrarily complex conjunctions and disjunctions nested within each clause. The conclusions are drawn if the premises are satisfied, making the rules purely inferential. Each rule thus states a single independent chunk of knowledge succinctly and states all necessary information explicitly in the premise [17]. The need to combine certainties is thus limited to the single existence of a given diagnostic rule. This greatly simplifies the certainty calculations of the rules' conclusions.

The only drawback to this approach is that evidence supporting a hypothesis is combined directly with the evidence negating that hypothesis into a simple metric for overall support. This makes it difficult to identify directly which support is either lacking or conflicting the given hypothesis. This drawback is generally encountered only if the KBS user wants an explanation of why particular fault hypotheses are not given.

The various possible clauses of our diagnostic rules are represented as *fuzzy sets*. Fuzzy sets [18] are used to define concepts or categories that have inherent vagueness and degree. Fuzzy set theory is thus based on the idea of gradual membership in a set. A fuzzy set is characterized by a membership function that relates a degree of membership on a scale of 0.0 to 1.0 to inherent properties of the objects. Fuzzy logic is thus being perceived increasingly as a way of handling continuous valued variables rather than uncertainty [19].

A fuzzy set proposition is a statement that asserts a value of a fuzzy variable. The fuzzy set proposition formulas allow conjunctions, disjunctions, and negation in the antecedents of the SV&PFA diagnostic rules. The relevant formulas are [13]:

For conjunction: $CF(A \text{ and } B) = min(CF(A), CF(B))$
For disjunction: $CF(A \text{ or } B) \ = max(CF(A), CF(B))$
For negation: $CF(\text{not } A) \ = 1.0 - CF(A)$

The intuitive appeal of these formulas is strong [20]. The probability of two events cannot be better than that of the least likely, and the probability of either of two events cannot be worse than the most likely. Furthermore, the formula for negation is the standard one used in probability theory; it is based on an observation that the probabilities of A and not A add up to 1.0.

4.4 MOME FUZZY LOGIC ALGORITHM

As described, MOME is based on an evaluation of process models of normal operation. The pattern of residuals that result is interpreted to determine any underlying assumptions that are currently not holding. More generally, such a process model residual may be represented generically as follows:

$$\text{residual} = f(x_1, x_2, \ldots, x_n) \tag{4.1}$$

where $x_1, x_2, \ldots, x_n$ are process assumption variables, that is, measured variables and unmeasured parameters that define the normal operating state of a process at any given moment, and f is a function of those variables that computes, for example, a balance of energy or mass in a control volume.

Because the sensors that measure process variables may not be totally accurate or provide exact readings, because the process models may not be perfect models of the relationship between the process variables, and because random process operations' perturbations may occur, it is observed empirically that the residuals are not always zero, although they are usually close to zero when the process is operating normally. The mathematical model describing normal process operation used by our program requires that all residuals be zero, on average, when a monitored process is operating normally. Therefore, a calculation is made from historical plant data of the average value of each primary model residual, and that normal average offset value is subtracted from the corresponding process model residual calculation.

In practice, then, each function f will behave like a statistical random variable having a mean value β and a standard deviation σ. The mean value $\beta = \beta_0 \rho$ and the standard deviation $\sigma = \sigma_0 \rho$, where β_0 and σ_0 are constants and ρ is either 1 or a process variable that is the definitive measure of the production level of the process being monitored. Usually, β is just a constant value, but sometimes it is the product of a constant times a process variable whose value determines the level of production at which the process is operating.

The generic process model residual above can be replaced with a primary model residual r, defined as

$$r = f(x_1, x_2, \ldots, x_n) - \beta \tag{4.2}$$

which has a mean value of zero and a standard deviation of σ. The equation defining r is referred to herein as a *primary process model residual*. The program examines the values of such (adjusted) process model residuals and, among other things, infers from the pattern of deviations from zero which sensors or other parts of the process may be faulty.

In more generic terms, if a process engineer provides the formula $f(\cdots)$ as the formula for a process model residual under ideal conditions, the formula *mean* for the average of $f(\cdots)$ over time based on historical plant data, and the formula *sigma* for the standard deviation of $f(\cdots)$ over time, the program generates the primary process model residual:

$$r = f(\cdots) - mean \tag{4.3}$$

which has the property that the average of r is expected to be zero. In reality, any formula that can be expressed in the mathematical language of

FALCONEER™ IV is allowed. The formula *sigma* is not used in evaluating the primary model residuals but is used to calculate certainty factors, as discussed below.

As discussed in Chapter 2, primary process model residuals are distinguished from certain linearly dependent process model residuals generated automatically by our program. Such additional models are referred herein to as *secondary process model residuals* and are computed as follows. Suppose that r_1 and r_2 are primary process model residuals and both contain a common variable v. If both primary models are linear functions of v, they may be combined algebraically to remove the terms containing v. An easier approach is to use equation (2.5). The standard deviation for this secondary residual process model can be computed directly from its two-parent model standard deviations using equation (2.6).

4.4.1 Single-Fault Fuzzy Logic Diagnostic Rule

This implementation of the MOME diagnostic strategy computes certainty factors to identify faults and/or validate underlying assumptions. As used below, a *fault* is a pair consisting of a process variable v and a direction d and is designated as $<v, d>$. The value of d can be either *high* or *low*, which are in turn defined by

$$high = 1 \tag{4.4}$$

$$low = -1 \tag{4.5}$$

Any number of different functions for computing certainty factors from residuals may be used. A commonly used function, the *Gaussian function*, is defined as

$$Gauss(r, sigma) = \exp\left[\frac{-(r/sigma)^2}{2}\right] \tag{4.6}$$

where *sigma* is the standard deviation of the model residual r.

When this program monitors a process, it reads real-time sensor data, computes the associated primary and/or secondary model residual values and their standard deviations, and then calculates three certainty factors for each residual value, as needed. Let r be one of the residual values and let *sigma* be its standard deviation. Residual r is expected to be zero, but often it is not. If it is only a little off from zero, the program considers it to be satisfactory, but the farther from zero it gets, the less confidence there is that it is satisfactory.

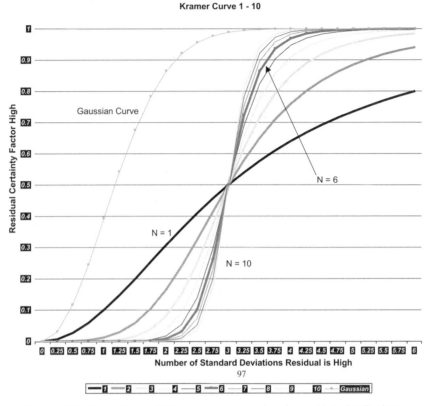

Figure 4.1 Gaussian and Kramer curve certainty factor high distributions.

Using the Gaussian function, for example, the certainty factor for r being satisfactory is calculated as follows:

$$cf(r, sat) = Gauss(r, sigma) \tag{4.7}$$

If r is much greater than zero, it is considered to be high, that is, higher than it is supposed to be. The certainty factor for r being high is represented as[4]

$$cf(r, high) = \begin{cases} 1.0 - Gauss(r, sigma) & \text{if } r > 0.0 \\ 0.0 & \text{otherwise} \end{cases} \tag{4.8}$$

---

[4]Equation (4.8) is plotted in Figure 4.1.

Similarly, if r is much less than zero, it is considered to be low. The certainty factor for r being low is represented as:

$$cf(r, low) = \begin{cases} 1.0 - Gauss(r, sigma) & \text{if } r < 0.0 \\ 0.0 & \text{otherwise} \end{cases} \quad (4.9)$$

As stated above, for purposes of calculating certainty factors for faults, $<v, d>$ signifies a fault; that is, v is a process variable and d is a direction, either *high* or *low*. To compute the certainty factor that a fault is present, the certainty factors for the primary and/or secondary model residuals are examined to find evidence for the fault. If r is a residual, r provides evidence for fault $<v, d>$ when it has deviated from zero in a direction that is consistent with variable v deviating in the direction d. For example, if $\partial r/\partial v$ is greater than zero, both v and r can be expected to go high (or low) at the same time. If, however, $\partial r/\partial v$ is less than zero, then v and r can deviate in opposite directions. The certainty factor for r in the appropriate direction is then the strength to which r can provide evidence for the fault. One strong piece of evidence for the fault is enough to conclude strongly that the fault is present unless there is also strong evidence that it is not present.

The evidence for fault $<v, d>$ is this set of certainty factors for all relevant primary model residuals:

evidence-for-fault $(<v, d>)$

$$= \left\{ cf\left[r, sign\left(\frac{\partial r}{\partial v} \right) d \right] \middle| \frac{\partial r}{\partial v} \neq 0 \text{ and } r \text{ is a primary residual} \right\} \quad (4.10)$$

In some applications, the certainty factor for any residual, primary or secondary, may be included in this set.[5] The strength of the evidence for the fault is the maximum of the values in this set.

Similarly, if a residual deviates in the opposite direction from what is expected when the fault is present, that deviation is evidence against the fault being present. The evidence against fault $<v, d>$ is this set of certainty factors for all relevant residuals:

$$\textit{evidence-against-fault } (<v, d>) = \left\{ cf\left[r, -sign\left(\frac{\partial r}{\partial v} \right) d \right] \middle| \frac{\partial r}{\partial v} \neq 0 \right\}$$

$$(4.11)$$

---

[5]By default, as discussed in Chapter 3, FALCONEER™ IV currently relies only on primary model residuals for calculating evidence-for-fault.

Certainty factors for both primary and secondary model residuals may be included in this set. The strength of the evidence against the fault is the maximum of the values in this set. If that value is subtracted from 1, the strength to which this evidence is consistent with the fault being present is determined.

An additional consideration is significant in evaluating a certainty factor for a fault. Some residuals are not functions of v and so are not expected to deviate from zero when the fault $<v, d>$ is present. The secondary model residual that was formed by eliminating v from two primary model residuals is such a residual. It is relevant to evaluating the presence of the fault, so this secondary model residual is expected to have a high certainty factor of being satisfactory when the fault involves v. Also, if two primary model residuals were combined to generate a secondary model residual process model by eliminating some variable other than v, and one of these primary model residuals is a function of v but the other is not, it is expected that the primary model residual that is not a function of v is satisfactory. This primary model residual is considered relevant to the fault as well.

Some primary process model residuals may not be functions of v and are not combined with any models that are. These are considered to be not relevant to the fault $<v, d>$. Another fault can be present and cause them to deviate from zero, but this will not affect the assessment for fault $<v, d>$. This allows a diagnosis of the presence of several single faults that happen not to interact with each other (i.e., a multiple-fault diagnosis[6]). In addition, r may be a function of v, but at the moment, $\partial r / \partial v = 0$. The neutral evidence for fault $<v, d>$ is this set of certainty factors for all relevant residuals:

$$neutral\text{-}evidence\ (<v, d>)$$
$$= \{cf(r, sat) | r \text{ is relevant as neutral evidence for } v\} \qquad (4.12)$$

The strength of this evidence is the minimum of the set because if any one of the residuals that are supposed to be satisfactory is in fact high or low, that weakens the evidence for the fault $<v, d>$.

The certainty factors in these three sets, *evidence-for-fault, neutral-evidence*, and *evidence-against-fault*, are considered to be fuzzy logic values

---

[6]Such situations are referred to as *non-interactive multiple-fault situations* [21]. These situations are defined as those for which the intersection of the sets of relevant primary and secondary models required to diagnose their member single-fault situations results in the empty set. These diagnoses are directly possible with equation (4.13) because independent inference pathways exist through the SV&PFA diagnostic rules based on MOME. This is an extremely important artifact of this methodology because multiple operating events (combinations of fault and/or nonfault situations) frequently occur and must be properly handled by the fault analyzer for it to be considered robust in actual applications.

[13] and are combined using a common interpretation of fuzzy AND as the minimum function, fuzzy OR as the maximum function, and fuzzy NOT as the complement function (1.0 minus the value of its argument). For finite sets, the quantifier SOME is just the OR of the values in the set, so it is equivalent to taking the maximum of the set. Similarly, for finite sets, the quantifier ALL is just the AND of all the values in the set, so it is equivalent to taking the minimum of the set. Putting this all together, the certainty factor for fault $<v, d>$ is defined in this fuzzy logic implementation as

$$cf(<v, d>) = SOME(evidence\text{-}for\text{-}fault(<v, d>))AND$$
$$ALL(neutral\text{-}evidence(<v, d>))AND$$
$$NOT(SOME(evidence\text{-}against\text{-}fault(<v, d>))) (4.13)$$

If this value is above the alert threshold, the corresponding fault is displayed as a possible current fault situation to the process operators.

Regarding the display of a fault $<v, d>$: If v is a measured variable, the sensor value for that variable was substituted for the variable in computing all the primary and secondary model residual values. If $d = high$, a conclusion is drawn that the sensor reading is higher than the true value for that process variable. If $d = low$, a conclusion is drawn that the sensor reading is lower than the true value for that process variable. In either case, a conclusion is drawn that the sensor is at fault. If $cf(<v, d>)$ is about zero for both cases, $d = high$ and $d = low$, the sensor reading has been validated (i.e., the sensor is correctly measuring its associated process variable).

In the case of an unmeasured variable v, such as a process leak, a high certainty factor for $<v, low>$ means that the assumed value of v, which can be viewed as the reading from a virtual sensor, is low compared to the actual value. To display a conclusion about the actual value of the unmeasured variable, the program displays a message that v is high in this case (i.e., there is a positive leak out of the process). Similarly, if the certainty factor for $<v, high>$ is high, it displays a message about v being low. If neither of these cases apply, a conclusion is drawn that the real value of v is about equal to its assumed normal and/or extreme value (i.e., the associated fault is not present and the parameter's value is currently indeed its normal and/or extreme value).

4.4.2 Multiple-Fault Fuzzy Logic Diagnostic Rule

In the vast majority of process systems, multiple-fault situations normally occur very much less frequently than do single-fault situations. Nevertheless, the ability to diagnose multiple-fault situations is important because (1) some major process disasters have occurred as a consequence of a series of two or more concurrent process fault situations, and (2) the patterns of diagnostic evidence (i.e., the SV&PFA diagnostic rules) used to diagnose single-fault situations may sometimes misdiagnose multiple-fault situations. The latter reason is equally true for diagnostic rules used by human troubleshooters, and in fact represents a contributing factor in some major process disasters [22]. Furthermore, certain types of multiple-fault situations may occur much more frequently than others. Kramer [3] describes three classes of such multiple faults as follows (1) faults causing other faults (called *induced failures*), (2) latent faults which are not detectable until additional faults occur, and (3) intentional operation in the presence of one or more faults, with the sudden occurrence of an additional fault.

The fuzzy logic diagnostic rule described above [i.e., equation (4.13)] was derived to detect and distinguish single process operating events. These events may be either fault or nonfault situations. Thus, the methodology described above is inherently a strategy for diagnosing any and all single assumption variable deviations that cause detectable operating events, not just fault situations per se. This is an important distinction, because not all of the potential assumption variables required in the primary models and performance equations for describing the normal operation of a typical process system correspond to actual process fault situations. For example, unsteady-state operation, unusually low production rates, changeover of a feed supply, normal process shutdown, and so on, are all examples of assumption variable deviations that may generate diagnostic evidence. Moreover, such process operating events may occur very frequently, much more so than the vast majority of single-fault situations and in many instances may accompany single-fault situations. Consequently, if the potential occurrences of multiple assumption variable deviations are not properly accounted for by the fault analyzer, the correctness of its diagnoses would always be questionable. Even if a fault analyzer cannot directly diagnose all such multiple assumption deviations (i.e., all possible combinations of process faults and nonfault operating events), it still needs to ensure that its diagnostic rules will not misdiagnose those situations. This is an extension of the conservative philosophy underlying MOME to multiple assumption deviations.

Throughout the following discussion, the term *multiple-fault situation* actually refers to any situation of multiple assumption variable deviations, whether or not all of those deviations correspond to actual process fault situations. This will not affect the generality of this discussion whatsoever.

The fuzzy logic diagnostic rule defined above [equation (4.13)] may be generalized to sets of faults by redefining what counts as evidence for the set, evidence against the set, and what counts as neutral evidence. An inference may be drawn that a set of faults is present when no subset of them may be inferred to be present. In particular, this means that there must be at least one primary residual value for each fault deviating in the direction that the fault can cause.[7] This leads to the following general fuzzy rule of this methodology:

Let

$$fault\text{-}set = \{<v_1, d_1>, \ldots, <v_n, d_n>\} \qquad (4.14)$$

---

[7]These situations are referred to as *interactive multiple-fault situations* [21]. These situations are defined as those for which the intersection of the sets of relevant primary and secondary models required to diagnose their member single-fault situations does not result in the empty set. There are three different types of such situations, identified further as fully discernible, partially discernible, and indiscernible.

Fully discernible interactive multiple-fault situations are those diagnosed directly by equation (4.15) and by lower-resolution single-fault situations diagnosed by equation (4.13). Furthermore, if identified by equation (4.15), it is possible that equation (4.13) may also identify erroneous but equally plausible single-fault situations (see Example 3.3).

Partially discernible interactive multiple fault situations are those that are not diagnosed directly by equation (4.15) but equation (4.13) identifies one or more plausible single-fault situation hypotheses. Although they are incorrect, these single-fault diagnoses are still all considered equally plausible explanations of the current process state. Those types of misdiagnoses represent a major potential problem, however, because the unannounced actual fault situations could be extremely hazardous, requiring that immediate corrective actions be taken by the process operators. It is also possible that the choice of the appropriate corrective actions will depend on whether or not any of the unannounced fault situations are currently occurring. This is also true about indiscernible interactive multiple-fault situations but with the following caveat. If the fault analyzer only announces an incorrect (although completely logically plausible) single-fault diagnosis, the operators may be led to believe that the fault being announced is the only one that could possibly be occurring. This may inadvertently cause them to take inappropriate corrective actions, subsequently causing the actual fault situation to become more dangerous. This is the one Achilles' heel of the MOME diagnostic strategy. It is something to be aware of always when using the program in actual process applications.

Indiscernible interactive multiple faults are those that cannot be diagnosed by either equation (4.13) or (4.15). This occurs because there are no unique primary models that depend on only one of the assumptions violated by that multiple-fault situation. The fault analyzer would report no fault diagnoses emerging from either equation (4.13) or (4.15) in this situation. This is considered acceptable performance by it because to classify a fault situation correctly the fault analyzer must either get the correct diagnosis or remain silent. As discussed in Appendix B, this behavior by the fault analyzer is considered conservative because it is only venturing diagnoses of which it is highly certain. Remaining silent about the actual multiple-fault situation in these cases thus does no harm as long as abnormal process operating situations are independently noticed and acted on by process operators in a timely fashion. Consequently, such misdiagnoses are considered acceptable performance by the fault analyzer.

Then

$$cf(fault\text{-}set) = SOME(evidence\text{-}for\text{-}fault(< v_1, d_1 >)) \, AND$$

.

.

.

$$SOME(evidence\text{-}for\text{-}fault(< v_n, d_n >)) \, AND$$
$$ALL(neutral\text{-}evidence(fault\text{-}set)) \, AND$$
$$NOT(SOME(evidence\text{-}against(fault\text{-}set))) \quad (4.15)$$

The component evidence sets are defined as follows:

- The set *evidence-for-fault*($<v_i, d_i>$) is the set of $cf(r, \, sign(\partial r/\partial v_i)d_i)$ such that r is a primary residual and $(\partial r/\partial v_i) \neq 0$. It is the same set that was used for single faults.
- The set *neutral evidence*(*fault-set*) is the set of $cf(r, \, sat)$ such that r is relevant to one or more of the faults in *fault-set* as neutral evidence and such that $(\partial r/\partial v) = 0$ for all the variables v in *fault-set*.
- The set *evidence-against*(*fault-set*) is the set of $cf(r, \, -sign(\partial r/\partial v)d)$ such that $(\partial r/\partial v) \neq 0$ for at least one variable v in *fault-set* provided that $-sign(\partial r/\partial v)d$ has the same value for all such variables v in *fault-set*. (In other words, all the faults that can influence r must influence it in the same direction; if two faults can influence r to deviate in opposite directions, we can learn nothing about *fault-set* from that residual.)[8]

The generalized rule for multiple-fault situations [equation (4.15)] is currently limited in FALCONEER™ IV to compute only certainty factors for potential pairs of interactive multiple faults. Typically, diagnostic resolution of potential multiple fault situations is almost always considerably much lower (see Example 3.4) than the typical diagnostic resolution of potential single-fault situations. There are typically many more possible combinations of potential multiple-fault effects on the current process operating state to consider.

A final issue regarding multiple-fault analysis in our methodology must be discussed. MOME as implemented above creates structured diagnostic rules relying on the minimum number of relevant primary models to diagnose specific faults, leaving all other remaining primary models to be considered irrelevant to those diagnoses. One advantage of this structure is that it creates

---

[8]Consequently, because the multiple fault influences the given model both high and low, that model can be in any of its three possible states whenever that multiple-fault situation occurs, the actual state depending on the actual relative magnitudes of the underlying member faults.

many independent inference pathways to its various assumption variable deviations. Such independent pathways allow all noninteractive multiple-fault situations to be diagnosed directly with the same diagnostic rule as that used to diagnose single-fault situations [equation (4.13)]. The only potential problem with the diagnostic rules for single-fault situations is that they could misdiagnose an interactive multiple fault that can generate that same pattern of violated and satisfied primary and secondary models as that generated by a single fault (as discussed, these fully discernible interactive multiple-fault situations are always possible; see Example 3.3). There are two possible approaches to handling such potential misdiagnoses.

One approach to handling this potential problem is simply to ignore it. In most situations, this can usually be done safely because the likelihood of a particular multiple-fault situation occurring is usually relatively small in comparison to that of a particular single-fault situation. In addition, requiring that the magnitudes of those possible assumption variable deviations which comprise that particular multiple-fault situation have particular ratios to each other makes it an even more unlikely candidate fault hypothesis. Thus, it is fairly safe to assume that a multiple-fault situation creating such diagnostic evidence will almost never occur in the target process system.

Making such a judgment goes beyond those made when defining the fault analyzer's intended scope. Those judgments eliminated entire classes of higher-level multiple faults, not just particular instances of multiple-fault situations actually contained within the intended scope (see Example 3.4). The decision to rule out unlikely fault hypotheses in the manner described above is based on the principle of parsimony, or Occam's razor [23]. This principle states that the simplest explanation consistent with all of the diagnostic evidence available should be preferred over all other plausible explanations that are considerably less likely. A similar principle, the *no-miracle rule* [24], is used in fault tree analysis to ignore possible but highly unlikely events that could interfere with the normal propagation of a basic event into a top-level event.

Another way around this potential problem of misdiagnoses is to give equal plausibility to both the single-fault hypothesis diagnosed by equation (4.13) and multiple-fault hypotheses diagnosed by equation (4.15). However, since multiple faults are normally much less likely to occur than single-fault situations, the majority of fault situations diagnosed would actually turn out to be the single-fault situation. Consequently, exhaustively listing all of the other plausible, but more highly improbable multiple-fault hypotheses could turn out to be potentially confusing to the process operators (see Example 3.3).

As currently implemented, by default, FALCONEER^{TM} IV follows the philosophy that all plausible fault diagnoses, either single and/or multiple, should be presented equally to the process operators. It is up to them to decide how much credence should be placed in lower-probability events than in more common fault occurrences. This directly combats the human operator limitations of "mind-set" [22][9] by enumerating all plausible explanations of abnormal process behavior that gets updated continuously as additional diagnostic evidence becomes available over time. It should be noted that multiple-fault analysis based on equation (4.15) can easily be disabled as a user control option when running FALCONEER™ IV applications if the users wish to do so. This in effect lets the program user employ Occam's razor directly to the subsequent fault analysis, but doing so eliminates the possibility of diagnosing all multiple-fault situations except noninteractive and fully discernible situations diagnosed directly at lower levels of diagnostic resolution using equation (4.13).

4.5   CERTAINTY FACTOR CALCULATION REVIEW

The algorithm described above to calculate certainty factors is based on a mixture of standard statistical inference and fuzzy logic [13]. This allows a fault hypothesis to have membership in all three possible outcome states (i.e., low, satisfied, or high) simultaneously, with the highest of those three values indicating which is the most likely state. Certainty factor calculations occur via the following three-step procedure:[10]

Step 1. Convert all primary and secondary models' residuals to certainty factors for each possible model state (i. e., low, satisfied, high).

First, convert all primary and secondary models' residuals into three certainty factors for each model (low, satisfied, high) [i.e., equations (4.7), (4.8), and (4.9)]. The certainty factors for low, satisfied, and high model states can be determined from the normal Gaussian distribution of those residuals or other continuous functions.[11] FALCONEER™ IV allows users to use one

---

[9]This limitation is also known as *cognitive lockup* or *cognitive narrowing* [25], *tunnel vision* [26], and the *point of no return* [27] (see Chapter 1).

[10]Operating in such a fashion allows the following three attributes to exist in the resulting certainty calculation [28]: (1) *locality*: each rule should be separate from all other rules; (2) *detachment*: once determined, a certainty can be understood independent of how it was derived; and (3) *truth functionality*: the truth of complex sentences can be computed from the truth of their components.

[11]In practice, however, certainty factors based on the Gaussian distribution lead to poor fault analyzer performance results when analyzing even moderately noisy sensor data.

such continuous function recommended by Kramer [3]. It is computed with the following formula:

$$\text{Kramer}(r, sigma) = \frac{1.0}{\{1.0 + [r/(3 \cdot sigma)]\}^N} \tag{4.16}$$

These values are stored for the primary and secondary models for **low**, **satisfied**, and **high** states as defined for Kramer's distribution by equations similar to (4.7), (4.8), and (4.9) for Gaussian distributions. On the moderately to extremely noisy data normally encountered in most actual applications of FALCONEER^{TM} IV, we recommend using Kramer's distribution with $N = 6$.[12]

Step 2. Evaluate every SV&PFA rule defined by the generalized MOME fuzzy logic diagnostic rules [equations (4.13) and (4.15)] with the certainty factors above for each of the three possible assumption variable states (i.e., low, satisfied, high) and determine which of the various distinct assumption variable states has a certainty factor that surpasses the display limit chosen for alerting the process operators.

All certainty factors for all fault hypotheses calculated from the fuzzy logic diagnostic rules [equations (4.13) and (4.15)] using residual certainty factors calculated in this manner can range in value from 0.0 to 1.0, with the sum of the certainty factors for the low, satisfied, and high states of a given assumption variable deviation equal to 1.0. A certainty factor of 0.0 indicates that either there is not at least one key piece of supporting evidence present for that associated hypothesis and/or that there is at least one piece of direct evidence against that hypothesis. A certainty factor of 1.0 indicates that the associated hypothesis has the minimum necessary supporting evidence and no direct evidence against it. These certainty factors are recalculated on every program cycle (as specified by the user-selected analysis interval) and compared against the alert limits to determine if sufficient credence in those associated hypotheses is present to warrant an alert to the users. Since these certainty factors are updated continuously after every vector of process data is received and analyzed, it helps combat the human limitations of vigilance decrement [29], mind-set [22], and cognitive overload [25,30,31] described in Chapter 1. FALCONEER^{TM} IV charts the certainty factors for all models and associated measured assumption variables, allowing users to monitor incipient failures as they develop over time.

---

[12]The Kramer curve formulas for various values of *N* using the high residual case are plotted in Figure 4.1.

Step 3. Sort the various alerted fault hypotheses in descending order for presentation to the process operators.

A simple sort routine is used to create both a current single-fault list and a current multiple-fault list of descending certainty factors. Note that because of the inclusive OR situations existing between potential hypotheses in the SV&PFA diagnostic rules, these lists could contain hypotheses that turn out to be erroneous. Nonetheless, those lists are also guaranteed to almost always contain all the correct hypotheses. This amounts to a trade-off between diagnostic resolution and diagnostic sensitivity by the fault analyzer and is the best that can be accomplished when all possible levels of resolution are included. Using all possible levels of resolution at varying levels of diagnostic sensitivity in this manner is desirable because many incipient faults will be "caught" earlier (i.e., identified at lower magnitudes of error or rates of occurrence), allowing corrective or counteractive measures to be performed in a more timely manner.

The procedure above for calculating certainty factors associated with the possible SV&PFA hypotheses accommodates the extensive usage of AND, inclusive and exclusive OR, and NOT logic in its fuzzy logic SV&PFA diagnostic rules (corresponding to their equivalent minimums, maximums, and complements). This logic results directly from applying non-Boolean reasoning techniques to MOME. All available evidence is included in each analysis, and only those hypotheses supported by that evidence are presented to the process operators. This minimizes possible nuisance alerts while such fault analyzers perform *intelligent supervision* of their target process systems.

4.6 MOME FUZZY LOGIC ALGORITHM SUMMARY

The method of minimal evidence employs model-based reasoning to logically deduce the cause or causes of abnormal process behavior. It does so with the least amount of diagnostic evidence necessary to uniquely diagnose the various possible fault situations. Moreover, the resulting fault analyzer always makes competent diagnoses at the best resolution and highest sensitivity possible for the given magnitude of the fault occurring.

One of the two key features of this diagnostic strategy is that it uses default reasoning to derive unique fault hypotheses. This is done in the following way. All but one of the fault hypotheses being supported by some of the diagnostic evidence contained within a given pattern of diagnostic evidence is shown systematically to be implausible by some of the other evidence also contained within that pattern. Thus, by default, the remaining fault hypothesis is the only plausible explanation of the full pattern of diagnostic evidence.

Using default reasoning in this manner allows this diagnostic strategy to base each fault diagnosis on the least amount of diagnostic evidence possible.

The other key feature of this strategy is in the way in which the various patterns of diagnostic evidence are selected. Their selection ensures that the resulting fault hypotheses are both the most sensitive and at the highest resolution possible for uniquely diagnosing their respective current fault situations. This is done by fully utilizing all the information contained within the available diagnostic evidence, that is, estimates of the fault magnitude for linear assumption variable deviations inherent in the primary models' residuals violated.

Combining the use of certainty factor calculations and fuzzy logic reasoning together with these techniques for deriving optimal diagnostic rules leads to the creation of structured knowledge bases. This greatly simplifies the inference procedure required during the fault analysis to a sequential calculation of the certainty factors associated with the various possible process fault situations. Consequently, the most plausible process fault hypotheses will be those with the largest associated certainty factors.

REFERENCES

1. Chester, D. L., L. Daniels, R. J. Fickelscherer, and D. H. Lenz, U.S. Patent 7,451,003, "Method and System of Monitoring, Sensor Validation and Predictive Fault Analysis," 2008.
2. Fickelscherer, R. J., D. H. Lenz, and D. L. Chester, "Fuzzy Logic Clarifies Operations," *InTech*, October 2005, pp. 53–57.
3. Kramer, M. A., "Malfunction Diagnosis Using Quantitative Models and Non-Boolean Reasoning in Expert Systems," *AIChE Journal*, Vol. 33, 1987, pp. 130–147.
4. Shell, P. S., *Expert Systems: A Practical Introduction*, Wiley, New York, 1985, p. 86.
5. Pearl, J., "On Evidential Reasoning in a Hierarchy of Hypotheses," *Artificial Intelligence*, Vol. 28, 1986, pp. 9–15.
6. Glymour, C., "Independence Assumptions and Bayesian Updating," *Artificial Intelligence*, Vol. 25, 1985, pp. 95–99.
7. Heckerman, D., "Probabilistic Interpretations for Mycin's Certainty Factors," in *Uncertainty in Artificial Intelligence*, ed. by L. N. Kanal and J. F. Lemmer, North-Holland, New York, 1986, pp. 167–196.
8. Horvitz, E., and D. Heckerman, "The Inconsistent Use of Measures of Certainty in Artificial Intelligence Research," in *Uncertainty in Artificial Intelligence*, ed. by L. N. Kanal and J. F. Lemmer, North-Holland, New York, 1986, pp. 137–151.

9. Paass, G., "Consistent Evaluation of Uncertain Reasoning Systems," presented at *the Sixth International Workshop on Expert Systems and Their Applications*, Avignon, France, April 1986.

10. Chandrasekaran, B., and M. C. Tanner, "Uncertainty Handling in Expert Systems: Uniform vs. Task-Specific Formalisms," in *Uncertainty in Artificial Intelligence*, ed. by L. N. Kanal and J. F. Lemmer, North-Holland, New York, 1986, pp. 102–113.

11. Buchanan, B., and E. H. Shortliffe, eds., *Rule-Based Expert Systems*, Addison-Wesley, Reading, MA, 1984.

12. Shafer, G., *A Mathematical Theory of Evidence*, Princeton University Press, Princeton, NJ, 1976.

13. Zadeh, L. A., "Fuzzy Logic," *Computer*, Vol. 21, No. 4, April 1988, pp. 83–93.

14. Gashnig, J., P. Klahr, H. Pople, E. Shortliffe, and A. Terry, "Evaluations of Expert Systems: Issues and Case Studies," in *Building Expert Systems*, ed. by F. Hayes-Roth, D. A. Waterman, and D. B. Lenat, Addison-Wesley, Reading, MA, 1983, p. 265.

15. Davis, R., and D. B. Lenat, *Knowledge-Based Systems in Artificial Intelligence*, McGraw-Hill, New York, 1982, p. 412.

16. Kline, P. J., and S. B. Dolins, "Choosing Architectures for Expert Systems," *Texas Instruments Inc. CCSC Technical Report 85-01-001*, Dallas, TX, October 1985, p. 66.

17. Davis, R., and D. B. Lenat, *Knowledge-Based Systems in Artificial Intelligence*, McGraw-Hill, New York, 1982, pp. 244–245.

18. Stefik, M., *Introduction to Knowledge Systems*, Morgan Kaufmann Publishers, San Francisco, CA, 1995, p. 531.

19. Russell, S., and P. Norvig, *Artificial Intelligence: A Modern Approach*, Prentice Hall, Upper Saddle River, NJ, 1995, p. 467.

20. Sell, P. S., *Expert Systems: A Practical Introduction*, Wiley, New York, 1985, pp. 88–89.

21. Fickelscherer, R. J., *Automated Process Fault Analysis*, Ph.D. dissertation, University of Delaware, Newark, DE, 1990.

22. Kletz, T. A., *What Went Wrong?: Case Histories of Process Plant Disasters*, Gulf Publishing, Houston, TX, 1985.

23. Reggia, J. A., D. S. Nau, and P. Y. Wang, "Diagnostic Expert Systems Based on a Set Covering Model," in *Developments in Expert Systems*, ed. by M. J. Combs, Academic Press, London, 1984, pp. 35–58.

24. Rolland, H. E., and B. Moriarity, "Fault Tree Analysis," in *System Safety Engineering and Management*, Wiley, New York, 1983, pp. 215–271.

25. Sheridan, T. B., "Understanding Human Error and Aiding Human Diagnostic Behavior in Nuclear Power Plants," in *Human Detection and Diagnosis of System Failures*, ed. by J. Rasmussen and W. B. Rouse, Plenum Press, New York, 1981, pp. 19–35.

26. Lees, F. P., "Process Computer Alarm and Disturbance Analysis: Review of the State of the Art," *Computers and Chemical Engineering*, Vol. 7, No. 6, 1983, pp. 669–694.

27. Rasmussen, J., "Models of Mental Strategies in Process Plant Diagnosis," in *Human Detection and Diagnosis of System Failures*, ed. by J. Rasmussen and W. B. Rouse, Plenum Press, New York, 1981, pp. 251–258.

28. Russell, S. J., and P. Norwig, *Artificial Intelligence: A Modern Approach*, Prentice Hall, Upper Saddle River, NJ, 1995, p. 415.

29. Eberts, R. E., "Cognitive Skills and Process Control," *Chemical Engineering Progress*, December 1985, pp. 30–34.

30. Bailey, S. J., "From Desktop to Plant Floor: A CRT Is the Control Operator's Window on the Process," *Control Engineering*, Vol. 31, No. 6, June 1984, pp. 86–90.

31. Fortin, D. A., T. B. Rooney, and H. Bristol, "Of Christmas Trees and Sweaty Palms," in *Proceedings of the Ninth Annual Advanced Control Conference*, West Lafayette, IN, 1983, pp. 49–54.

5

METHOD OF MINIMAL EVIDENCE: CRITERIA FOR SHREWDLY DISTRIBUTING FAULT ANALYZERS AND STRATEGIC PROCESS SENSOR PLACEMENT

5.1 OVERVIEW

In this chapter, the criteria for distributing fault analyzers are described. We show that they can be distributed effectively into networks that roughly parallel currently existing distributed process control systems. Also described in this chapter are the criteria used to locate additional process sensors strategically so that fault analysis can be performed more effectively. These additional sensor measurements are typically selected to improve a fault analyzer's diagnostic sensitivity and/or its diagnostic resolution for particular fault situations.

5.2 CRITERIA FOR SHREWDLY DISTRIBUTING PROCESS FAULT ANALYZERS

In this section we discuss the issues surrounding distributed process fault analyzers.

Optimal Automated Process Fault Analysis, First Edition.
Richard J. Fickelscherer and Daniel L. Chester.
© 2013 John Wiley & Sons, Inc. Published 2013 by John Wiley & Sons, Inc.

5.2.1 Introduction

The first step in developing a fault analyzer is to specify its target process system. This amounts to defining a control volume around all the process components of interest. Doing so implicitly defines the entire set of the potential process fault situations and nonfault events within the fault analyzer's intended scope. It also determines explicitly on which process sensor measurements the fault analyzer can base its analysis.

More important, specifying the target process system determines which process engineers and operators need to be consulted during the fault analyzer's development. This is extremely important because they are the most knowledgeable about the normal (and abnormal) operating behavior of the target process system. They are also the ones whom the fault analyzer is intended to aid. If the resulting fault analyzer does not meet their needs, the chances that it will be used will be greatly reduced. Moreover, their trust in the fault analyzer's advice will be greatly enhanced if they have a firm understanding of the reasoning used by the fault analyzer. The best way for them to gain such an understanding is by active involvement in the fault analyzer's development. This was true in the development of both the original FALCON and various subsequent FALCONEER^{TM} IV knowledge-based system (KBS) applications. Having these personnel actively involved during its entire development will also improve the likelihood that they will both be willing and able to maintain the fault analyzer as required.

5.2.2 Practical Limitations on Target Process System Size

One practical limitation placed on the size of the target process system is related to the resulting fault analyzer's ability to operate in real time. Regardless of the particular inference procedure being used, the fault analyzer needs to be sufficiently responsive to search the entire diagnostic knowledge base correctly in real time[1]; this in turn allows the fault analysis to be performed continuously online.

Fault analysis based on the method of minimal evidence (MOME) can be accomplished by an extremely simple inference procedure; specifically, only a sequential calculation of the certainty associated with the various fault hypotheses is required, with a simple sort of the results. This is possible because the logical inference required to derive fault hypotheses is compiled directly into the structured knowledge base. Consequently, it is possible to

---

[1]Real-time analysis is considered to occur if the fault analyzer collects its required sensor data, performs all of its fault analysis with the underlying SV&PFA diagnostic knowledge, and reports its results before the next analysis cycle begins.

perform it efficiently in one of the traditional programming languages (i.e., Visual Basic was used to create the FALCONEER™ IV KBS) rather than in one of the relatively more exotic artificial intelligence languages (i.e., VAX Common Lisp was used to create the original FALCON KBS). This is highly advantageous because typically most plant personnel are much more adept at maintaining software written in the traditional programming languages.

In addition, most traditional programming languages are compatible with the computer operating systems currently used by and networked into modern distributed control systems. Consequently, the need to maintain either customized and/or exotic computer software or hardware systems is minimized, which eliminates a major obstacle to the use of artificial intelligence programs in real time online process control applications [1].

However, fault analyzers using simple inference procedures still have limitations. Because an exhaustive search is required to ensure that all plausible fault hypotheses are made, an upper limit will always exist on the number of diagnostic rules that can be searched in real time. For a given data sampling rate, this places an upper limit on how many processing units can be included within the target process system.[2]

To get around this potential limitation, a variety of inference strategies that operate in "as needed" mode have been suggested [3–8]. These strategies eliminate the need to perform exhaustive searches at the most detailed level of modeling abstraction, thus allowing the fault analyzer to continuously analyze a larger subsection of the total process system in real time. However, such strategies require that the process system be modeled at multiple levels of abstraction, something that may be difficult to do properly. These strategies also have the various overheads associated with using artificial intelligence software in industrial settings. Bench test studies have shown that procedural programs (e.g., those written in Visual Basic, C, Fortran) operate about one to two orders of magnitude faster than comparable rule-based programs [9]. This indicates that complex inference routines implemented in artificial intelligence languages need to be able to improve the efficiency of the reasoning strategy by at least a similar amount.

Even if the entire process system could be analyzed in real time by a single knowledge-based system, it is probably not the best way to perform fault analysis. The diagnostic messages emanating from such a fault analyzer would be received by a central agent. These messages would then have to be sorted and properly communicated to the control rooms in charge of the malfunctioning process units. This would represent additional overhead costs in both program development and in real-time operation. Maintaining such complicated programs would probably also be much more difficult.

---

[2] A strategy to decompose large process systems based on qualitative process fault analysis has been reported [2]. Qualitative process fault analysis is described in Section A.6.

5.2.3 Distributed Fault Analyzers

One way to alleviate these problems is to analyze the overall process system with a number of small, distributed fault analyzers. The distributed fault analyzers would analyze process subsystems corresponding roughly to those currently being monitored by the distributed control rooms in the processing plant. This division of the process system is also ideal for leveraging the knowledge held by the process experts. Having such a distributed network of fault analyzers will probably also make it possible to operate each one on relatively inexpensive microcomputers attached directly to the distributed control system.

Diagnostic knowledge bases created with MOME can be distributed effectively by grouping the primary models according to the assumption variable deviations that affect those constraints. These models are grouped according to the assumption variable deviations that can violate the largest number of primary models. For most process systems, these assumption deviations will correspond to interlock activations on the major feed flows. Consequently, the optimal control volume will be defined by the location of the process variables that can activate interlock shutdowns of the major feed flows.[3] Thus, the process variables used to define the optimal control volumes will tend to parallel those monitored by the distributed control rooms throughout the processing plant.

Unfortunately, defining a target process system in such a way directly reduces the sensor information available to each of its associated fault analyzers. Consequently, having a distributed network of fault analyzers will adversely affect the diagnostic resolution of these fault analyzers for at least some potential fault situations. This occurs because the process models used in those faults' corresponding diagnostic rules would require sensor data that are being collected outside a given control volume and thus could not be evaluated directly.

A way to counteract this problem is to have the various target process systems overlap one another. Even if this were done properly, though, the resulting fault analyzers would still have suboptimal diagnostic resolution for some potential fault situations. However, if the overlapping target process systems were chosen shrewdly, the suboptimal resolution of neighboring fault analyzers within the distributed network could be completely resolved by logical inference.

For example, suppose that one fault analyzer concluded that either fault situation A or B was occurring and that a neighboring system's fault analyzer

---

[3]This was one of the criteria used by DuPont to select the LTC recycle loop as the target process system for the original FALCON system (see Appendix B).

concluded that fault situation B or C was occurring. For this situation it would be possible to conclude that fault situation B was actually occurring.[4] Now suppose that the first fault analyzer again concluded that either fault situation A or B was occurring, but the second fault analyzer did not make any diagnoses. For this situation, the diagnosis would depend on the estimated fault magnitude of fault situation B. If it were lower than the diagnostic sensitivity of the second fault analyzer, the supervisory fault analyzer would conclude that either fault situation A or B was occurring. Otherwise, it would conclude that fault situation A was occurring. In this way, the supervisory agent would be used to coordinate the diagnoses of the various lower-level fault analyzers within the distributed network. For most fault situations, this would ensure that the same level of diagnostic resolution obtained simply by using a single-fault analyzer could also be obtained with the distributed network of fault analyzers.

5.3   CRITERIA FOR STRATEGIC PROCESS SENSOR PLACEMENT

Currently, process plants are instrumented so that their operation can be more easily monitored and controlled. Normally, little thought is given to the impact this has on an operator's ability to perform process fault management effectively.[5] Systematic approaches are not normally used to analyze proposed configurations of process instrumentation for performing optimal fault analysis. In the past, poorly designed configurations have been a contributing cause of many major process accidents [11], including the near meltdown of the nuclear reactor at the Three Mile Island power plant [12]. Diagnostic methods to deal directly with less than ideal instrumentation have been proposed [13].

Systematic approaches to process fault analysis, such as MOME, can be used directly to analyze the impact of process instrumentation on the ability to perform fault analysis effectively.[6] In fact, such analyses were used twice during the development of the original FALCON KBS to examine the instrumentation of the adipic acid system. The first analysis performed became the basis for a recommendation that DuPont collect 15 additional

[4]This assumes that the principle of parsimony described in Chapter 4 (i.e., Occam's Razor) is being employed.
[5]Optimal instrumentation location in process plants in the past was typically based on cost-minimization methods rather than optimizing potential process fault analysis [10]. This situation is starting to change throughout the CPI.
[6]An algorithm for locating sensors to improve diagnostic resolution for sign-directed graph diagnostic strategy (described in Section A.5) is given by Raghuraj et al. [14].

process measurements (seven pairs of controller outputs and setpoints and one thermocouple reading). This recommendation was accepted. The second analysis was performed to anticipate the effects that an impending process system modification would have on the FALCON system's performance. This formed the basis of a recommendation that DuPont leave in place a pressure sensor that it had originally planned to remove. This recommendation was also accepted. Such analyses were also used in the development of the FALCONEER™ IV KBSs to recommend to FMC that two additional flowmeters be added to their electrolytic sodium persulfate system and that a flowmeter and three thermocouples be added to their electrolytic liquid ammonium persulfate system.

In general, process instrumentation, whether unique or redundant, should be added only if it will appreciably improve the fault analyzer's diagnostic sensitivity and/or its diagnostic resolution for a particular fault situation. Moreover, since adding and maintaining instrumentation is expensive, it must be justified by tangible improvements in process safety and/or productivity. Obviously, fault situations that could have potentially severe consequences on either safety or productivity would probably qualify as those that should have all the instrumentation necessary for performing effective fault analysis.

REFERENCES

1. Basta, B., "Process-Control Techniques Make Rapid Advances," *Chemical Engineering*, Vol. 94, No. 15, 1987, pp. 22–25.
2. Lee, G., and E. S. Yoon, "A Process Decomposition Strategy for Qualitative Fault Diagnosis of Large-Scale Processes," *Industrial and Engineering Chemistry Research*, Vol. 40, No. 11, 2001, pp. 2474–2484.
3. Moore, R. L., L. B. Hawkinson, C. G. Knickerbocker, and L. M. Churchman, "A Real-Time Expert System for Process Control," in *Proceedings of the First Conference on Artificial Intelligence*, Denver, CO, 1984, pp. 528–540.
4. Moore, R. L., and M. A. Kramer, "Expert Systems in On-Line Process Control," in *Proceedings of the Third International Conference on Chemical Process Control*, Asilomac, CA (A Cache Publication), Elsevier, New York, 1986, pp. 839–867.
5. Moore, R. L., B. Hawkinson, M. Levin, A. G. Hofmann, B. L. Matthews, and M. H. David, "Expert Systems Methodology for Real-Time Process Control," in *Proceeding of the 10th World Congress on Automatic Control*, IFAC, Munich, Germany, July 1987, Vol. 6, pp. 274–281.
6. Shum, S. K., J. F. Davis, W. F. Punch, and B. Chandrasedaran, "An Expert System Approach to Malfunction Diagnosis in Chemical Plants," *Computers and Chemical Engineering*, Vol. 12, 1988, pp. 27–36.

7. Rich, S. H., and V. Venkatasubramanian, "Model-Based Reasoning in Diagnostic Expert Systems for Chemical Process Plants," *Computer and Chememical Engineering*, Vol. 11, No. 2, 1987, pp. 111–122.

8. Laffey, T. J., P. A. Cox, J. L. Schmidt, S. M. Kao, and J. Y. Read, "Real-Time Knowledge Based Systems," *AI Magazine*, Spring 1988, pp. 27–45.

9. Brownston, L., *Programming Expert Systems in OPS5: An Introduction to Rule-Based Programming*, Addison-Wesley, Reading, MA, 1985.

10. Bagajewicz, M., "A Review of Techniques for Instrumentation Design and Upgrade in Process Plants," *Canadian Journal of Chemical Engineering*, Vol. 80, No. 1, 2002, pp. 3–16.

11. Kletz, T. A., *What Went Wrong?: Case Histories of Process Plant Disasters*, Gulf Publishing, Houston, TX, 1985.

12. Lees, F. P., "Process Computer Alarm and Disturbance Analysis: Review of the State of the Art," *Computer and Chememical Engineering*, Vol. 7, No. 6, 1983, pp. 669–694.

13. Howell, J., "Model-Based Fault Detection in Information Poor Plants," *Automatia*, Vol. 30, No. 6, 1994, pp. 929–943.

14. Raghuraj, R., M. Bhushan, and R. Rengaswamy, "Locating Sensors in Complex Chemical Plants Based on Fault Diagnostic Observability Criteria," *AIChE Journal*, Vol. 45, No. 2, 1999, pp. 310–322.

6

VIRTUAL SPC ANALYSIS AND ITS ROUTINE USE IN FALCONEER™ IV

6.1 OVERVIEW

This chapter provides an overview of FALCONEER™ IV's virtual statistical process control (virtual SPC) analysis. This technology should allow anyone with the capability of continuously monitoring particular process sensor readings or calculating performance equations (a.k.a., key performance indicators or soft sensors) the ability to determine whether or not they are under control. Its analysis is based on exponentially weighted moving averages of those readings or performance equation calculations over time. This analysis allows for out-of-control sensors or calculated variables to be flagged more quickly and at levels that may allow the process operators to intercede with the control actions necessary to mitigate the underlying process problems without unduly disrupting process operations. It is considered virtual because the analysis is done automatically without the need for operators to collect and chart process sensor readings.

The normal standard deviation and mean of each measured variable for which virtual statistical process control will be performed needs to be evaluated to configure those variables properly. As described in Chapter 2 for calculating the necessary primary models statistics, this evaluation is performed automatically in FALCONEER™ IV also on individual data tags with the same three to six months' worth of normal process data already collected and

Optimal Automated Process Fault Analysis, First Edition.
Richard J. Fickelscherer and Daniel L. Chester.
© 2013 John Wiley & Sons, Inc. Published 2013 by John Wiley & Sons, Inc.

loaded for those primary models. Next we describe how FALCONEER™ IV uses these individual tag standard deviations and means for both controlled and uncontrolled measured variables and performance equation calculated variables.

6.2 INTRODUCTION

Statistical process control (SPC) is a tool used to assess whether a process is currently under or out of control. Various techniques exist for doing this analysis, depending on the nature of the process being monitored. In continuous processes (as opposed to the manufacture of discrete, individual units), process data collected at a particular moment in time are not completely independent of its previous data. This phenomenon is referred to as *autocorrelation* among the data. SPC techniques used to deal with autocorrelation in the data revolve around calculations of *exponentially weighted moving averages* (EWMAs). These calculations cancel the effects of autocorrelation and allow small but statistically significant shifts in an observed or calculated variable's value to be readily detected. Calculating EWMAs is the method used in our virtual SPC software module to determine if a particular sensor reading or performance equation (a.k.a., key performance indicator or soft sensor) calculation is in or out of control.

In FALCONEER™ IV, there are two types of measured variables and one type of calculated variable to contend with when configuring the particular virtual SPC analysis to be performed. Measured variables must be defined either as controlled variables or uncontrolled variables. All variables calculated must be defined as performance equation variables. A different EWMA interpretation is performed for each of these types of variable, depending on how the particular variable was configured for virtual SPC analysis in its corresponding tag editor. Each of these possible interpretations is described below.

6.3 EWMA CALCULATIONS AND SPECIFIC VIRTUAL SPC ANALYSIS CONFIGURATIONS

The formula for computing an EWMA is [1]

$$\text{EWMA}_t = \lambda Y_t + (1 - \lambda) \cdot \text{EWMA}_{t-1} \tag{6.1}$$

where Y_t is the current value of the monitored variable, EWMA_t the currently calculated EWMA, EWMA_{t-1} the EWMA calculated previously, and λ the weighting factor. λ needs to be determined for each particular variable, but good results are typically obtained for values of 0.1 to 0.2. The sampling time also needs to be determined, but good results occur if it is greater than the process time constant of the process.[1] The results of the EWMA calculation are interpreted differently depending on the nature of the variable being monitored (either controlled, uncontrolled, or performance equation variables) and the way it was configured for virtual SPC analysis. All possible configurations are discussed below.

6.3.1 Controlled Variables

For controlled variables that have been configured without a manual mean or manual control limit (its standard configuration), the upper and lower control limits (UCL and LCL) on their EWMA charts are given by

$$\text{UCL} = \mu_Y + 3.0\sigma_Y \cdot \text{SQRT}\left(\frac{\lambda}{2.0 - \lambda}\right) \tag{6.2}$$

$$\text{LCL} = \mu_Y - 3.0\sigma_Y \cdot \text{SQRT}\left(\frac{\lambda}{2.0 - \lambda}\right) \tag{6.3}$$

where μ_Y is the controlled variable's (i.e., Y_t's) setpoint value and σ_Y is its standard deviation of Y_t about that value. This configuration thus allows μ_Y to change if the controlled variable's associated setpoint changes during the analysis. For this configuration, the current value of the EWMA (i.e., EWMA_t) is compared to these UCL and LCL to determine its virtual SPC alarm status. From their definitions, both the UCL and the LCL are equidistant from the current setpoint value. Red virtual SPC alarms occur whenever the current EWMA goes beyond either the UCL or the LCL, and yellow virtual SPC alarms occur whenever it goes beyond two-thirds of the interval between the setpoint and these two limits.

A variation on this standard configuration for controlled variables is to define a manual mean. This makes μ_Y constant in the calculation of the UCL and the LCL. These limits thus do not change if the actual setpoint for the specific control variable changes. Controlled variables configured in this manner still have their current calculated EWMA values (i.e., EWMA_t) values compared to these stationary UCL and LCL to determine their virtual SPC alarm status. Again, from their definition, both the UCL and the LCL

---

[1]The process time constant for a system is the summation of the transport delay times and three times the residence times of any tanks in that system.

are equidistant from the manual mean and the red and yellow alarms occur exactly as they do when the setpoint is used as the mean.

A final variation on the standard configuration for controlled variables is to define both a manual mean and manual control limits. This makes μ_Y constant and makes both the UCL and the LCL constant but not necessarily equidistant from the manual mean. Controlled variables configured in this manner still have their current calculated EWMA values (i.e., $EWMA_t$) compared to these UCL and LCL to determine their virtual SPC alarm status as above. However, if the manual mean is chosen to be different from the actual mean of the UCL and LCL chosen, such configurations allow the virtual SPC yellow alarm intervals to be skewed either high or low, correspondingly.

6.3.2 Uncontrolled Variables and Performance Equation Variables

For uncontrolled variables and performance equation variables, the standard configuration is also to define them without setting a manual mean or manual control limits. For these two types of variables, since the EWMA statistic can be viewed as a one-step-ahead forecast of the process mean for the next period, the one-step-ahead predication error, $\varepsilon_1(t)$, can be used directly to establish the upper and lower control limits on their EWMA charts. The absolute value of the predication error, $\Delta(t)$, is the method used in this case [2]:

$$\varepsilon_1(t) = Y_t - EWMA_{t-1} \tag{6.4}$$

$$\Delta(t) = \alpha|\varepsilon_1(t)| + (1-\alpha)\Delta(t-1) \tag{6.5}$$

where α is the first-order filter constant and $\Delta(t-1)$ is the absolute value of the prediction error for the last time period $t-1$.

Since $\sigma \approx 1.25\,\Delta(t)$ for a normal distribution, the UCL and LCL used for uncontrolled variables and performance equation variables are predictive by definition:

$$UCL_{t+1} = EWMA_t + 3.75\Delta(t) \tag{6.6}$$

$$LCL_{t+1} = EWMA_t - 3.75\Delta(t) \tag{6.7}$$

The choice of α controls how much of the historical process data is used in estimating the standard deviation of the prediction error. Larger values of α put more weight on recent data, while smaller values of α put more weight on older data. Good results are typically obtained when the value is 0.01 to 0.1.

For both uncontrolled and performance equation variables defined without a manual mean or manual control limits (neither constant nor fixed distance), their current values (i.e., Y_t) are compared to both the UCL and the LCL based on the past value of the EWMA (i.e., the value of $EWMA_{t-1}$) to determine their virtual SPC alarm status. From their definitions, both the UCL and LCL are equidistant from the past EWMA value. Red virtual SPC alarms occur whenever the current value (i.e., Y_t) goes beyond either the UCL or LCL, and yellow virtual SPC alarms occur whenever it goes beyond two-thirds of the interval between the past EWMA value and these two limits.

A common situation that can occur when using this virtual SPC analysis configuration of uncontrolled and performance equation variables is that if the current values (i.e., Y_t) do not change over time (i.e., remain constant long enough), the values of the UCL and LCL both eventually converge to that value and the virtual SPC alarms sound. Although this is a direct method of detecting stuck sensors, it also causes many false alarms when the UCL and LCL shrink down over time, and then the current value (i.e., Y_t) does change rather insignificantly. For this reason, other possible configurations can be employed to help ensure that only significant changes in the monitored variable are brought to the user's attention. These are described below.

FALCONEER^{TM} IV also allows Uncontrolled and performance equation variables the ability to set the UCL and LCL values to constant fixed limits. This is referred to as *precontrol*. For both types of variables, the current EWMA calculated value is compared directly against these limits to determine the virtual SPC alarm status as above. When configuring these limits, it is also necessary to specify a manual mean for that variable. If the manual mean is chosen to be different than the actual mean of the constant UCL and LCL chosen, such configurations skew the virtual SPC yellow alarm intervals either high or low, correspondingly. In practice, precontrol limits are used mostly to set upper and lower alert limits on the interpretation of calculation results of performance equation variables (a.k.a., key performance indicators). It can, however, also be used for uncontrolled variables which are determined not to have control issues at the automatically calculated limits, making it possible to specify a normal range of operation for those variables based on safety or other concerns. This flexibility in defining alert limits allows users to better tune the performance of the virtual SPC analysis on actual process data for the various variables being monitored.

Other virtual SPC analysis configurations of uncontrolled and performance equation variables are also possible. One such configuration is referred to as *fixed distance*. In this configuration, if a manual mean is not set for this variable, the variable's UCL and LCL are defined to be a specified number of standard deviations from the current EWMA calculated value (i.e., $EWMA_t$). The variable's current value (i.e., Y_t) is then compared against those limits to determine its virtual SPC alarm status. If a manual mean is set for this variable, the variable's UCL and LCL are defined to be a specified number of standard deviations from this manual mean. In this configuration, the variable's current EWMA calculated value (i.e., $EWMA_t$) is then compared against those limits. (The latter configuration is equivalent to the precontrol configuration, with the exception that the limits are always equidistant from the manual mean chosen and that those constant limits get calculated from the variable's standard deviation.)

A final variation on the virtual SPC analysis configuration of these two types of variables is simply to set a manual mean on the variable. In this case, their current values (i.e., Y_t) are compared to both the UCL and the LCL based on the past value of the EWMA (i.e., $EWMA_{t-1}$) to determine their virtual SPC alarm status. From their definitions, both the UCL and the LCL are equidistant from the past EWMA value. Red virtual SPC alarms occur whenever the current value (i.e., Y_t) goes beyond either the UCL or the LCL. However, now the yellow virtual SPC alarms occur whenever it goes beyond two-thirds of the interval between the manual mean value and these two limits. Depending on the value of this manual mean and the past value of the EWMA, this may skew the yellow alarm region compared to that which would occur if this manual mean were not set. Although allowed as a potential virtual SPC analysis configuration, in practice it does not usually lead to improvements in determining out-of-control situations over the case where the manual mean is not set. This configuration is therefore not recommended in actual applications.

As described above, depending on how they are configured, uncontrolled and performance equation variables have their virtual SPC alarm status based on either their current values (Y_t's) or their current EWMA values (i.e., $EWMA_t$). In contrast, control variables always have this status based on their current EWMA values (i.e., $EWMA_t$). In either case, exceeding either the UCL or the LCL generates a red alert (out of control); exceeding two-thirds of the interval defined between either the UCL or the LCL and the chosen mean generates a yellow alert (going out of control); else it is a green alert (in control). Again, this analysis will always be performed periodically at a frequency that can be defined to be different for each variable being monitored. Choosing this frequency properly for each variable will depend on consideration of the various time constants inherent in the actual process system and the process data sampling rate. This determination is necessary for each variable being monitored since continuous process data are autocorrelated and thus require sufficient time between samples before unique information is forthcoming by the virtual SPC analysis. When configured correctly, this analysis provides a powerful tool for proactively determining the current status of process operations so that timely corrective actions can be taken to help optimize those operations.

6.4 VIRTUAL SPC ALARM TRIGGER SUMMARY

As described in Section 6.3, the particular virtual SPC alarms that can occur for a particular type of variable depend on how the virtual SPC analysis is configured for that variable. Depending on these configurations, either the variable's CV (i.e., current value) or current EWMA calculated value (updated at the interval specified in that variable's virtual SPC analysis configuration) is compared against the high and low yellow and red virtual SPC alarm limits. These limits, again depending upon the type of variable and its associated configuration, can either be fixed or have changing values. The following list summarizes which of these possible values (either CV or EWMA) and corresponding limits (fixed or changing) is used for the different possible configurations. This corresponds to the value (either CV or EWMA) reported as the trigger for the virtual SPC alarms recorded in those reports.

Controlled Variables   These variables can be configured for virtual SPC analysis either as:

1. Standard EWMA (analysis occurs as described in Section 6.3.1, comparing EWMA with the UCL and LCL centered on the controlled variable's associated setpoint value).
2. Fixed mean (analysis occurs as described in Section 6.3.1, comparing EWMA with the UCL and LCL centered on the manual mean). *This configuration is not recommended if the actual setpoint of the controlled variable changes frequently.*
3. Fixed mean and fixed limits (analysis occurs as described in Section 6.3.1, comparing EWMA with the manual UCL and LCL possibly skewed on the manual mean).

Uncontrolled and Performance Equation Variables   These variables can be configured for virtual SPC analysis as one of the following:

4. Standard EWMA (analysis occurs as described in Section 6.3.2, comparing CV with the UCL and LCL centered on the prior EWMA's value). *This configuration is not recommended if the CV does not change significantly over time; it will, however, eventually flag any stuck sensors or constant calculations.*

5. Fixed mean (analysis occurs as described in Section 6.3.2, comparing CV with the UCL and LCL centered on the prior EWMA's value but yellow alarm limits centered on the fixed mean, so they may be skewed). *This configuration is not recommended; use standard EWMA instead.*

6. Fixed mean and fixed distance (analysis occurs as described in Section 6.3.2, comparing EWMA with the UCL and LCL centered on the manual mean by the number of specified standard deviations).

7. Fixed distance (analysis occurs as described in Section 6.3.2, comparing CV with the UCL and LCL equidistant from the current EWMA by number of specified standard deviations).

8. Fixed mean and fixed limits (precontrol) (analysis occurs as described in Section 6.3.2, comparing EWMA with the manual UCL and LCL with the yellow alarm limits possibly skewed on the chosen manual mean).

This variety of potential configurations for performing the virtual SPC analysis makes FALCONEER™ IV extremely flexible for auditing the current performance of process operations and determining whether or not all monitored variables are currently in or out of control.

6.5 VIRTUAL SPC ANALYSIS CONCLUSIONS

Using virtual SPC to monitor the status of measured variables and performance equation calculations is an additional tool for optimizing process operations. It brings to bear a type of analysis shown to be useful with autocorrelated data, hopefully giving the process operators both more timely and more meaningful alerts than those already given by their distributed control system. This will then allow them to respond with the appropriate control actions sooner.

REFERENCES

1. Montgomery, D. C., *Introduction to Statistical Quality Control*, Wiley, New York, 1991, pp. 299–309.

2. Montgomery, D. C., *Introduction to Statistical Quality Control*, Wiely, New York, 1991, pp. 341–351.

7

PROCESS STATE TRANSITION LOGIC AND ITS ROUTINE USE IN FALCONEER™ IV

7.1 TEMPORAL REASONING PHILOSOPHY

Temporal reasoning in both the original FALCON and the various FALCONEER™ IV knowledge-based systems (KBSs) is limited to adjacent time intervals. Sampling of all the process data occurs at a predetermined frequency. However, context within the KBSs changes only when it is appropriate. The logic of the KBSs is concerned with the previous state (the one that existed one data time sample ago), the current inferred state (the one determined from the current sampled data) and the forecasted next state (predicted from the preceding and current inferred states) when determining its context. This is similar to the temporal reasoning employed by the ventilator manager (VM) KBS [1] and greatly simplifies these programs' underlying temporal logic, which otherwise would require more elaborate representations of events and time.[1]

This problem—that the process state changes continuously over time—requires that the fault analyzer also change its prior presented plausible fault hypotheses continuously over time. This problem is widely known within the artificial intelligence community as the *frame problem* [2]. In such situations,

---

[1]See Appendix C for a detailed pseudocode example of a more complex temporal logic reasoning routine used by the original FALCONEER KBS at FMC.

Optimal Automated Process Fault Analysis, First Edition.
Richard J. Fickelscherer and Daniel L. Chester.
© 2013 John Wiley & Sons, Inc. Published 2013 by John Wiley & Sons, Inc.

new information causes an old conclusion to be withdrawn. Such a conclusion is said to be *defeated* [3].[2] The entire reasoning system is classified as *nonmonotonic*. Nonmonotonicity is not an unusual property of a reasoning system. On the contrary, most of the inferences done by real programs are *defeasible* [3]. For FALCONEER IV™ process fault analyzers, the reasoning employed during each analysis step is monotonic, meaning that the program firmly believes its conclusions at that time stamp. On the next analysis cycle, it completely forgets its previous conclusions and draws all new inferences. These new conclusions may or may not, depending entirely on what process operating state is currently occurring, confirm previous conclusions by the fault analyzer. The program's awareness is thus always focused on the current process state.

7.2 INTRODUCTION

Once the MOME diagnostic strategy was adopted as the desired model-based diagnostic methodology in the original FALCON project, a comprehensive attempt to formally verify the FALCON system's knowledge base began. A major problem that became more apparent during further testing with plant data concerned the primary models being used to detect and diagnose faults. There was a definite need to limit the conditions under which those various models were considered valid representations of the actual process behavior. It was evident that major process upsets, such as pump failures and interlock activations, invalidated many of the fundamental assumptions used during the development of those models. Consequently, it became necessary that the fault analyzer always determine the current operating state of the process before the diagnostic rules were applied. Such a determination was required to determine which of the specific modeling assumptions were still valid and thus which of the primary models would be appropriate in the fault analysis. To do this, process events such as process startups, shutdowns, and interlock activations needed to be monitored explicitly by the FALCON system.

This monitoring was accomplished by developing rules to detect the occurrence of such events and could subsequently determine the current operating state of the process system. The possible operating states became: (1) the process was being started up, (2) the process was in production; (3) the process was in production but was rapidly approaching an interlock activation; (4) the process had interlocked but production had ceased for less than 2 minutes,

---

[2]A conclusion that may later be defeated is *defeasible*.

and (5) the process had interlocked and production had ceased for more than 2 minutes. The transition between these process states was determined automatically by the fault analyzer, which then limited its analysis of the process data accordingly.

Two minutes after an interlock was chosen as the cutoff point for fault analysis because it represented the typical amount of time required for the effects of major process transients to subside after emergency shutdowns. After 2 minutes it is no longer possible to analyze current process data to determine the fault situation that caused the original emergency process shutdown. Moreover, after an interlock the majority of process models no longer made reliable predictions of the process behavior observed. It was therefore necessary to halt the diagnosis of all process faults that relied on those models, which turns out to be most of the target process fault situations monitored by the fault analyzer. Thus, 2 minutes after a process interlock, the FALCON system would go to sleep until the process system was started up again. While sleeping the fault analyzer would continue to monitor plant data, but it would be incapable of diagnosing process fault situations. When the process restarted, the FALCON system would automatically 'reawaken' (i.e., automatically begin analyzing again for all process fault situations). Messages informing the process operators of these state transitions were displayed automatically by the human–machine interface.

Adding this capability for automatically determining the current process state had several consequences. First, it allowed the FALCON system to be turned on when the process was either operating or shut down; the fault analyzer could automatically determine which state it was in. It also allowed the FALCON system to be run continuously, regardless of the plant state transitions that could occur. More important, adding this capability also logically structured the entire knowledge base according to the patterns of evidence contained within the various diagnostic rules. This reduced the fault analysis of the incoming process data to an ordered, sequential search through the entire set of possible process faults. This search sequence in effect constitutes a priority hierarchy between the various possible process fault situations. The logical structure of the fault priority hierarchy within the FALCON system's knowledge base is illustrated in Figure B.3. As discussed in Appendix B, discovering the rationale behind this hierarchy represented a major development in the MOME paradigm.

The enhancements caused a substantial increase in the size and complexity of the knowledge base. The upgraded version of the FALCON system's diagnostic knowledge base contained approximately 800 diagnostic rules (over 10,000 lines of Common LISP code). It was capable of detecting and diagnosing all of the 60 target process fault situations plus about 100 additional fault situations, most of which were malfunctions in the interlock shutdown system. It was also capable of detecting and diagnosing a few extremely dangerous fault situations that could occur during process startups.

A similar type of analysis was performed in the original FALCONEER KBS (pre-FALCONEER™ IV availability) developed for the FMC Corporation. The details underlying this analysis are described in pseudocode presented in Appendix C. The following describes how this analysis is currently performed in FALCONEER™ IV.

7.3 STATE IDENTIFICATION ANALYSIS CURRENTLY USED IN FALCONEER™ IV

State ID variables are by definition those used for determining the current operating state of the process (i.e., starting up, operating, shutting down, shutdown, etc.). These variables can be any measured system variable, continuous or discrete. When operating (i.e., actively analyzing actual process data), FALCONEER™ IV always first determines whether or not the process is operating within acceptable conditions. This is accomplished with the state identification (state ID) module. This module determines if the process is currently operating in a mode in which the FALCONEER™ IV analysis can derive meaningful results. FALCONEER™ IV is idle if the process is not operating. The program automatically begins its analysis of sensor measurements once process startup is complete. It continues this analysis until the process is shut down and then is idle again (i.e., passively monitoring process data) until the next process startup completes. This process condition monitoring suite thus runs continuously and adjusts its analysis appropriately to current process operations.

State ID variables are any sensor measurements used by the program to determine if the process is currently operating in production mode. These variables tend to monitor either key process feeds or process configuration states. They generate green alerts (used to denote normal operation) when the process is operating and each of these variables is either within its associated *standard operating conditions* (SOCs) or its normal state. If outside their SOCs but within their interlock limits, these variables generate yellow alerts. In either case, the program status is "In Production Mode" and "Not Expecting Interlocks." If outside their interlock limits or normal configuration, these variables generate red alerts and the other FALCONEER™ IV's analysis modules (SV&PFA and virtual SPC) are preempted from running. In this situation, the program status becomes "Not in Production Mode" and "Interlock Expected." It stays in this mode until none of the state ID variables are in red alert.

The state identification (state ID) analysis is always performed first when the process interval time since the last FALCONEER™ IV analysis cycle elapses. If all state ID variables and all other measured variables enabled in the state ID module are in their acceptable operating ranges (if continuous variables) or configurations (if discrete variables), the following analysis occurs, in this order:

1. Any non-state ID module variables which have BAD values are identified and all primary models that depend on them are disabled.
2. The SV&PFA analysis is performed with all still currently enabled primary models.
3. Any resulting variables that are determined to be either in red SV&PFA alarm or BAD values and are also used in any performance equations cause the results of those calculations to be classified as "Calculation Invalid" [i.e., performance equation calculated results are suspect because possibly invalid values of variables are being used in this calculation (or BAD)].[3]
4. the virtual SPC analysis is performed on all "GOOD" (i.e., not "BAD") measured variables and performance equations for which their individual virtual SPC analysis time interval has been surpassed.
5. The program actively monitors the system time to determine when the next FALCONEER™ IV analysis cycle should begin (i.e., whenever the analysis time interval defined for the particular process being monitored has been surpassed).

If any of the state ID variables or other measured variables enabled in the state ID module are not in their acceptable operating ranges (if continuous variables) or configurations (if discrete variables) (as indicated by their being in a red alarm state), steps 1 to 4 are skipped and the program goes directly to step 5. It continues in this mode until all variables in the state ID module are considered acceptable again (as indicated by them all being in either green or yellow alarm states). The possible state ID alarm states need further elaboration, depending on the type of variable being monitored (i.e., either discrete or continuous).

Discrete variables are defined directly as State ID variables with a defined normal value as the normal configuration status for that variable. This is its green alarm condition. Whenever it is not this normal value, the variable is in red alarm and preempts further FALCONEER™ IV analysis from occurring. There is thus no yellow alarm state associated with these state ID variables.

---

[3]See Example 3.2 for a discussion of this logic.

Continuous variables that can be defined in the state ID module analysis are either controlled or uncontrolled variables. The following six parameters need to be specified for either type defined in the state ID module:

1. Minimum measurement limit
2. Low interlock limit
3. Minimum standard operating condition limit
4. Maximum standard operating condition limit
5. High interlock limit
6. Maximum measurement limit

These parameters are used in the following manner to set the various state ID alarm states for these continuous variables:

Green. If the current value of the given variable is less than or equal to maximum standard operating condition limit and greater than or equal to the minimum standard operating condition limit, the variable is in the green state ID alarm state, and this variable does not preempt further FALCONEER™ IV analysis.

Yellow high. If the current value of the given variable is greater than the maximum standard operating condition limit but less than the High Interlock Limit, the variable is in the yellow high state ID alarm state, and this variable does not preempt further FALCONEER™ IV analysis.

Yellow low. If the current value of the given variable is less than the minimum standard operating condition limit but greater than the low interlock limit, the variable is in the yellow low-state ID alarm state, and this variable does not preempt further FALCONEER™ IV analysis.

Red high. If the current value of the given variable is greater than or equal to the high interlock limit but less than or equal to the maximum measurement limit, the variable is in the red high-state ID alarm state, and this variable does preempt further FALCONEER™ IV analysis.

Red low. If the current value of the given variable is less than or equal to the low interlock limit but greater than or equal to the minimum measurement limit, the variable is in the red low-state ID alarm state, and this variable does preempt further FALCONEER™ IV analysis.

Red high high. If the current value of the given variable is greater than the maximum measurement limit, the variable is in the red high-high-state ID alarm state, and this variable does preempt further FALCONEER™ IV analysis. These alarms are displayed to the user just as red high alarms are.

Red low low. If the current value of the given variable is less than the minimum measurement limit, the variable is in the red low-low-state ID alarm state, and this variable does preempt further FALCONEER™ IV analysis. These alarms are displayed to the user just as red low alarms are.

If any of these parameters are not defined for a particular variable, their values should be set equal to the next logical limit that is defined. This will eliminate some of these possible alarm levels. For example, if there is no maximum standard operating condition limit, this value should be set equal to the high interlock limit. This will eliminate the yellow high alarm state as a possible situation for that variable.

7.4 STATE IDENTIFICATION ANALYSIS SUMMARY

Again, the motivation for the state ID module is to preempt further analysis by FALCONEER™ IV when the process is not operating (e.g., shutdown) or is going through startup and has not lined out yet to normal operation (e.g., diverter valves in bypass configuration). It thus attempts to identify operating regimes where the corresponding SV&PFA and virtual SPC analysis is meaningful. Consequently, when none of the state ID variables or other variables enabled in the state ID module are red (either low, low low, high, or high high), FALCONEER™ IV further performs the appropriate SV&PFA and virtual SPC analysis and reports its results with the same alarm detail screens. One interacts with these subsequent alarms in the same manner as with the state ID alarms.

REFERENCES

1. Stefik, M., J. Aikins, R. Balzer, J. Benoit, L. Birnbaum, F. Hayes-Roth, and E. Sacerdoti, "The Architecture of Expert Systems," in *Building Expert Systems*, ed. by F. Hayes-Roth, D. A. Waterman, and D. B. Lenat, Addison-Wesley, Reading, MA, 1983, pp. 97–98.
2. Rich, E., *Artificial Intelligence*, McGraw-Hill, New York, 1983, pp. 176–180.
3. Charniak, E., and D. McDermott, *Introduction to Artificial Intelligence*, Addison-Wesley, Reading, MA, 1985, pp. 369–371.

8

CONCLUSIONS

8.1 OVERVIEW

This chapter summarizes our model-based diagnostic strategy, the method of minimal evidence (MOME) and its various advantages for optimally automating process fault analysis. It also describes a procedure for developing automated process fault analyzers using our program, FALCONEER™ IV, which is based directly on the fuzzy logic implementation of MOME described previously. This program allows fault analyzer development and maintenance to be highly cost effective in actual process system applications, allowing process engineers to create and maintain such programs themselves for their targeted process systems. The benefits derived from using FALCONEER™ IV on two real-world highly complex electrolytic process systems owned and operated by FMC Corporation in Tonawanda, New York are also discussed.

8.2 SUMMARY OF THE MOME DIAGNOSTIC STRATEGY

The chief advantage of MOME is that it provides a uniform framework for examining process models and their associated modeling assumptions. In this framework, each primary model represents the relationship that exists between its required modeling assumption variables during normal process operation.

Optimal Automated Process Fault Analysis, First Edition.
Richard J. Fickelscherer and Daniel L. Chester.
© 2013 John Wiley & Sons, Inc. Published 2013 by John Wiley & Sons, Inc.

Any significant residuals resulting from the evaluation of these models indicate directly that one or more of the associated modeling assumption variables are deviating significantly. As discussed, all modeling assumption variable deviations can be classified into one of three possible categories. These three categories are distinguished from one another by their effect on the residuals of the primary models they violate. This permits the derivation of standardized SV&PFA diagnostic rule formats, thus allowing the fault analyzer development to be completely systematic once the primary models are derived and deemed well-formulated.

As discussed, the major advantage of the MOME diagnostic strategy centers about the diagnostic rule format that it uses. The format has been chosen to ensure that the diagnostic knowledge bases will always perform competently (i.e., it will only make the correct fault diagnoses or not any diagnoses). The format also ensures that the resulting diagnostic knowledge bases can diagnose those faults with the best diagnostic sensitivity and diagnostic resolution possible for the current magnitude of the process operating event. Each diagnostic rule contains only the minimal amount of diagnostic evidence required to uniquely discriminate its associated fault situations from all other likely process operating events. As discussed, using just the minimal amount of diagnostic evidence in each diagnostic rule in this manner also allows many possible multiple-fault situations to be diagnosed directly. Also as discussed, this diagnostic methodology can be used directly to determine the strategic sensor placement for performing process fault analysis that further maximizes diagnostic sensitivity and/or resolution for particular process operating events. It can further be used to distribute fault analyzers shrewdly throughout a large process system.

8.3   FALCON, FALCONEER, AND FALCONEER™ IV ACTUAL KBS APPLICATION PERFORMANCE RESULTS

The original FALCON system was tested during its development with 260 simulated fault situations and 500 hours of selected process data containing 65 process operating events (e.g., emergency shutdowns, startups, production changes), including 13 actual process fault situations. During its off-line field test, the FALCON system monitored over 5000 continuous hours of process data in real time. These data included 22 process operating events, eight of which were actual process fault situations. The FALCON system was then tested online for three months. DuPont independently rated its performance at better than 95% correct responses during that test [1].

Although it preformed competently online, maintaining and improving FALCON's diagnostic knowledge base (written in Common Lisp–based data structures comprising 20 primary models and 800 + diagnostic rules) proved to be impractical for anyone other than the original developer (i.e., the University of Delaware) [2]. Being a research project, maintainability was not given as high a priority as was FALCON's performance with actual process data. From a research viewpoint, generalizing the underlying logic of this model-based diagnostic strategy was paramount, allowing future such development project activities to be as streamlined as possible. This effort led directly to the formulation of MOME [3, 4] It allows automated process fault analyzers to be developed systematically and eliminates the need for the exhaustive testing performed during the original FALCON project to demonstrate that the resulting programs are competent.

Development of the original FMC FALCONEER System [5, 6] proved to be a systematic application of MOME to their electrolytic sodium persulfate (ESP) process. Developing the 35 primary models and five performance equations that describe normal process operation, evaluating them with sufficient process data to determine their normal variances and offsets, and then hand-compiling the 15,000 + SV&PFA diagnostic rules consumed the vast majority of the development effort. It took an order of magnitude less development time and effort to create this program than to develop the original FALCON project program (1 person-year versus approximately 15 person-years), although the FMC ESP process was more than twice the scope of DuPont's adipic acid process.

Even this impressive improvement in the required development effort for the original FALCONEER system was still an order or two of magnitude higher in time requirements than those now necessary. This development time has been shortened remarkably by the automation of MOME in FALCONEER™ IV [7]. Specifically, all the development required for FMC's liquid ammonium persulfate (LAP) process was to derive about 30 primary models and five performance equations (requiring about two weeks of effort). All the requisite diagnostic logic required to evaluate our two fuzzy logic diagnostic rules is now derived directly from the underlying models of normal process operation. Consequently, all of the development and maintenance effort required to create such fault analyzers can be directed at the derivation of models of normal process operation (the declarative knowledge) in the configuration database. This constitutes the data structure input into FALCONEER™ IV's compiler. Since the set of all the primary models are linearly independent of each other, they can be added, improved, or deleted as need be and then the entire application recompiled to include these improvements. Consequently, with this compiler, the fault analyzer can now be improved incrementally with minimal effort as the process system's operating behavior becomes better understood or the process system changes, allowing the fault analyzer to evolve easily along with its associated process system. This, in turn, substantially reduces both the development and maintenance effort and subsequent costs of such programs. It effectively converts the much more difficult problem of automated process fault analysis into the much simpler and more directly tractable (and incrementally solvable) problem of process modeling.

Since it has been now fully automated, FALCONEER™ IV has been operating continuously online at FMC's Tonawanda, New York, active oxidants plant (on their ESP and LAP processes) since midsummer 2003. To date, FALCONEER™ IV has competently monitored its two associated process systems for possible process fault situations and other nonfault operating events as a diligent and relentless watchdog (i.e., it has exhibited high utility). A summary of the direct benefits derived from this continuous timely fault analysis was reported independently by FMC [8].

8.4 FALCONEER™ IV KBS APPLICATION PROJECT PROCEDURE

The FALCONEER™ IV process performance suite program is a real-time online process performance monitoring software tool that assists process operators and engineers by auditing current process operations to help ensure that they are optimal and to identify root causes of problems when those operations become abnormal. It accomplishes this by monitoring specific sensor measurements continuously, performing advanced calculations and analysis with them, and immediately giving advisory alerts when the process is not performing optimally or when actual processing problems occur. These alerts go beyond typical distributed control system alarms because FALCONEER™ IV uses engineering models of normal process operation and advanced statistical calculations in its analysis. From this analysis, FALCONEER™ IV is able to determine whether the sensor measurements monitored are correct, whether they are currently in control, whether other process faults are occurring, and whether the current overall process performance is optimal. The resulting timely alerts should help users more effectively and optimally to control and operate their process systems.

The advanced and systematic process auditing performed by FALCONEER™ IV while performing its associated state ID, SV&PFA, and virtual SPC analysis continuously extracts additional, highly useful information about current process operations from the current sensor measurements. The results of this analysis are available continuously in a variety of user-friendly formats so that these results can be reported automatically and immediately to the appropriate personnel (e.g., immediate intelligent alerts to operators, systematic e-mails to process engineers). This should enable corrective actions for improving process operations to be taken in a more timely, proactive manner, both reducing operating costs and improving process safety. These benefits occur continuously from more effectively extracting all of the relevant information currently available in the process data already being collected by the client's control system.

In general, the following steps are followed during a FALCONEER™ IV application project:

1. Identify a target process system for analysis.

2. Segment that target process logically according to its major unit operations or equipment, identifying them with unique process area names (or identifiers).

3. Identify all of the measured and unmeasured variables associated with each process area and determine their engineering measurement units.

4. Identify key measured variables that directly indicate the current process operating state and their associated limits. These are used directly by the state ID module to determine if further analysis of the measured variables collected during the current process state is warranted by FALCONEER™ IV.

5. Create as many primary models and performance equations (a.k.a., key performance indicators or soft sensors) as possible with the sensor measurements available in the target process.

6. Evaluate these primary models with sufficient normal process data (i.e., at least three to six months' worth) to determine typical model residual standard deviations and offsets as functions of production. These models and associated statistics are used directly by the fault analysis module to perform SV&PFA.

7. Evaluate the normal standard deviation and mean of each measured variable for which virtual statistical process control will be performed. These statistics are used directly by the virtual SPC module to do its associated analysis.

8. Determine upper and lower alert limits (if both are applicable) on performance equations (a.k.a., key performance indicators or soft sensors). These limits are used directly by the virtual SPC module to do its associated analysis.

9. Configure the FALCONEER™ IV application database (input primary models, performance equations, and associated calculated statistics) according to the configuration hierarchy suggested.

10. Choose the various FALCONEER™ IV time intervals to help ensure that its analysis is performed consistently and correctly.

11. Run the FALCONEER™ IV application with actual process data to perform the SV&PFA sensitivity analysis; make adjustments to FALCONEER™ IV's various SV&PFA tuning parameters as necessary.

12. Run the FALCONEER™ IV application with actual process data to perform the virtual SPC sensitivity analysis; make adjustments to FALCONEER™ IV's various virtual SPC tuning parameters as necessary.

13. Start using the FALCONEER™ IV application online in real time to help improve process operations.

A moderately complicated process system typically requires less than a person-month of effort to configure and fully validate a FALCONEER™ IV application.

8.5 OPTIMAL AUTOMATED PROCESS FAULT ANALYSIS CONCLUSIONS

Process data become valuable only when they are at the right place at the right time and there are mechanisms to interpret and use them properly [9]. Process data thus cannot be considered an asset unless correctly presented, analyzed, and converted into information [10]. "Moreover, we really don't want information; we want knowledge ... we want information analyzed, converted, and organized in a useful way" [11]. Using models to do this analysis and conversion and then performing sensor validation and proactive fault analysis (SV&PFA) based on those results in real time proves the assertion that "models are the means by which data can be converted to meaningful information" [12]. These models are based on a fundamental understanding of normal operating behavior of the given process system. This fundamental knowledge thus constitutes an unimpeachable source for direct generation of relevant information concerning normal and abnormal process behavior. Such information is immensely useful for logically inferring conclusions about the current process operating state. Performing this inference automatically based on MOME with FALCONEER™ IV after each update of process sensor data allows such fault analyzers to perform highly effective intelligent supervision of the daily operations of their associated process systems. "Thus, the whole business environment of process operations should become more rational. More information will be gathered, synthesized, and put into useful form more rapidly ... information organized in a useful way, after knowledge-able inferencing about its content" [11]. FALCONEER™ IV is thus a simple and easy-to-use tool now available to the process industries for proactively examination of live process information for any faults, so as to allow continuous monitoring of a given process for safer operation, better performance, and ultimately, higher efficiency and/or production levels. It has now become a very straightforward and cost-effective proposition to develop and maintain extremely competent automated process fault analyzers throughout the processing industries.

REFERENCES

1. Rowan, D. A., "Beyond FALCON: Industrial Applications of Knowledge-Based Systems," in *Proceedings of the International Federation of Automatic Control Symposium*, ed. by P. S. Dhurati, Newark, DE, 1992, pp. 215–217.

2. Rowan, D. A., and R. J. Taylor, "On-Line Fault Diagnosis: FALCON Project," in *Artificial Intelligence Handbook*, Vol. 2, Instrument Society of America, Research Triangle Park, NC, 1989, pp. 379–399.

3. Fickelscherer, R. J., *Automated Process Fault Analysis*, Ph.D. dissertation, University of Delaware, Newark, DE, 1990.

4. Fickelscherer, R. J., "A Generalized Approach to Model-Based Process Fault Analysis," in *Proceedings of the 2nd International Conference on Foundations of Computer-Aided Process Operations*, ed. by D. W. T. Rippin, J. C. Hale, and J. F. Davis, CACHE, Inc., Austin, TX, 1994, pp. 451–456.

5. Skotte, R., D. Lenz, R. Fickelscherer, W. An, D. LaphamIII, C. Lymburner, J. Kaylor, D. Baptiste, M. Pinsky, F. Gani, and S. B. Jørgensen, "Advanced Processss Control with Innovation for an Integrated Electrochemical Process," presented at the *AIChE Spring National Meeting*, Houston, TX, 2001.

6. Fickelscherer, R. J., D. H. Lenz, and D. L. Chester, "Intelligent Process Supervision via Automated Data Validation and Fault Analysis: Results of Actual CPI Applications," Paper 115d, presented at the *AIChE Spring National Meeting*, New Orleans, LA, 2003.

7. Fickelscherer, R. J., D. H. Lenz, and D. L. Chester, "Fuzzy Logic Clarifies Operations," *InTech*, October 2005, pp. 53–57.

8. Lymburner, C., J. Rovison, and W. An, "Battling Information Overload," *Control*, September 2006, pp. 95–99.

9. Harmon, P., and D. King, *Expert Systems: Artificial Intelligence in Business*, Wiley, New York, 1985, p. 2.

10. Kennedy, J. P., "Data Treatment," in *Proceedings of the 2nd International Conference on Foundations of Computer-Aided Process Operations*, ed. by D. W. T. Rippin, J. C. Hale, and J. F. Davis, CACHE, Inc., Austin, TX, 1994, pp. 21–44.

11. Harmon, P., and D. King, *Expert Systems: Artificial Intelligence in Business*, Wiley, New York, 1985, p. 212.

12. Kramer, M. A., and R. S. H. Mah, "Model-Based Monitoring," in *Foundations of Computer-Aided Process Operations II*, ed. by D. W. T. Rippin, J. C. Hale, and J. F. Davis, CACHE, Inc., Austin, TX, 1994, pp. 45–68.

A

VARIOUS DIAGNOSTIC STRATEGIES FOR AUTOMATING PROCESS FAULT ANALYSIS

A.1 INTRODUCTION

For a variety of reasons, past attempts at automating process fault analysis have not been very successful. They were much more expensive, in terms of the amount of effort, resources, and time than was originally planned. More important, the computer code of the resulting fault analyzers typically performed below expectations. This code also proved to be very difficult to maintain as the target process system evolved over time. Consequently, many projects were halted prematurely in the prototype stage of development. Fault analyzers that did get tested on actual process systems were generally abandoned shortly after those tests were completed. In this appendix we review some of these development projects, their diagnostic strategies, and the major problems encountered.

Currently, a diverse variety of logically workable diagnostic strategies exist for automating process fault analysis. Despite this, automated process fault analysis is still not widely used within the processing industries. The distinguishing logic of these various diagnostic strategies and their inherent shortcomings when used in actual applications are discussed. Specifically, the various diagnostic strategies reviewed are (1) fault tree analysis, (2) alarm analysis, (3) decision tables, (4) sign directed graphs, (5) methods using qualitative models, (6) methods using quantitative models, (7) methods using

Optimal Automated Process Fault Analysis, First Edition.
Richard J. Fickelscherer and Daniel L. Chester.
© 2013 John Wiley & Sons, Inc. Published 2013 by John Wiley & Sons, Inc.

artificial neural networks, and (8) methods using knowledge-based systems. The appendix concludes with our strategy recommendation for effectively performing real-world automated process fault analysis.

A.2 FAULT TREE ANALYSIS

A good discussion of fault tree analysis is that by Roland and Moriarity [1]. Briefly, it is a method that was developed in the early 1960s to analyze the safety of ICBM launch control systems. Currently, the method is used widely to analyze the safety of many complex systems, including chemical and nuclear process systems. Using Boolean logic, it describes the relationships that exist between the various possible basic events in a particular system (e.g., system component failures, human errors) and the final outcomes of those events (e.g., fires, explosions). The final outcomes are normally referred to as top events. Based on the relationships and the probabilities associated with the occurrence of the various basic events (referred to as those events' failure rates), fault trees can be used to estimate the probability associated with the occurrence of each top event [2,3].

Some of the largest drawbacks of this method are that constructing accurate fault trees (1) requires an extraordinarily detailed understanding of the process system being analyzed, (2) is a very time-consuming task, and (3) is a very prone to errors. To overcome the last two limitations, various methods to generate fault trees automatically have been proposed and tested [4–11]. Such methods are designed to make it easier to create and modify fault trees. However, these programs are all still in the prototype stage and are not ready for widespread use [9,10] because plant-specific estimates of failure probabilities are critical but in practice difficult to derive accurately.

Fault tree analysis is better suited for analyzing process system designs for potential safety problems than for diagnosing process faults in real time. Despite the basic mismatch in its underlying orientation, Teague and Powers[1] were first to present an algorithm that combines a priori estimates of failure rates with real-time process data to determine a quantitative basis for ordering the sequence in which potential fault hypotheses are examined. However, getting accurate failure rate data remains a major limitation of the approach.

Other drawbacks of fault tree analysis were discussed by Martin-Solis et al. [12]. These include the fact that (1) not all basic events can be observed, (2) difficulties are encountered when sensor measurements are incorrect,

[1]Teague, T. L., and G. J. Powers, "Diagnosis Procedures from Fault Tree Analysis," Carnegie-Mellon University, 1978 (unpublished manuscript).

(3) difficulties are encountered when devices fail in their normal operating positions, (4) traditional fault tree analysis assumes that time is not a factor in a fault's propagation through the process system, and (5) many of the fault situations generated by fault tree analysis are considered trivial and thus would have to be discounted in real-time applications. Despite these shortcomings, diagnostic strategies based on fault tree analysis continue to be proposed [7–11], most centered around knowledge-based system techniques [13–17].

A.3 ALARM ANALYSIS

Alarm analysis diagnoses faults by associating fault situations with the corresponding pattern and order of occurrence of process alarms in the target process system [18–23]. These relationships are used to create structured representations called alarm trees. Prime-cause alarms (i.e., those most closely associated with the actual process fault situations) are stored at the lower levels in the resulting alarm tree representations, while the other "effect alarms" are stored at higher levels. The alarm trees thus establish the priority relationships that exist between the various process alarms. The overall goal is to suppress irrelevant alarms, thereby helping the operators to focus their attention on the critical alarms. Alarm analysis thus attempts to intelligently reduce the amount of information that process operators have to interpret during major process upsets. It therefore addresses the problem of cognitive overload described in Chapter 1.

In practice, the method has not been too successful [20,23]. The reasons for this is that programs using alarm trees are costly to develop, are prone to errors, and are difficult to modify when the target process system is changed. To counteract some of these problems, algorithms have been proposed to generate alarm trees automatically [19,22]. Another problem is that unanticipated process system dynamics and process hardware problems can alter the patterns of the process alarms generated during a given process situation. This in turn may either suppress valid information or lead to confusing fault diagnoses. Consequently, it turns out that the most valuable fault trees are relatively small, typically connecting three or four alarms [21].

A.4 DECISION TABLES

Decision tables (also known as fault dictionaries) are derived from the cause-and-effect relationships that exist between process variables and process fault situations [24,25]. For a given fault situation, each process variable is assigned a value of low, normal, or high, depending on how that fault situation affects it. The key advantages of decision tables are that they are relatively easy to derive, implement, and understand. The information that they contain is also highly structured, which facilitates the examination of every possible combination of patterns. In turn, the patterns represented within the tables can be converted into simple Boolean logic expressions for computer implementation [24,26].

The disadvantages of this method are that (1) it does not take into account the possibility of sensor failure or process noise, (2) it does not take into account unsteady-state process operating conditions (such as the transients which almost always occur during a severe process fault situation), and (3) the tolerance limits on the variables may need to be different for the various fault situations. Also, as with alarm analysis, anything that alters the patterns of variable response will radically alter the resulting fault diagnosis. Furthermore, even if the patterns do not get distorted, the level of discrimination between the various process fault situations is typically not very good. Consequently, because of their limited usefulness, in actual applications decision tables have been coupled with other diagnostic strategies [20].

A.5 SIGN-DIRECTED GRAPHS

Sign-directed graphs follow the same logic of decision tables but are a more succinct method for representing all the possible patterns that can result from each process fault situation [27–35]. Nodes in these diagrams represent the process variables observed, while the arcs in these diagrams represent the causal relationships between the variables. The fundamental premise underlying this approach is that the origin node of a fault must be linked in the graph to all of the consequences of that fault that have been observed. As with decision trees, the advantages of this method are that it is relatively easy to use, implement, and understand.

The original limitations of this approach were that (1) only single faults could be diagnosed, (2) a fault could not cause a given variable to change sign in more than one direction, (3) the diagrams were inherently ambiguous (i.e., had poor diagnostic resolution), and (4) the algorithms proposed were not computationally efficient, making real-time operation difficult. The latter limitation was addressed by Kramer and Palowitch [30] by compiling these diagrams directly into Boolean expressions. Updated and improved algorithms are now available [32–35], with the methodologies allowing multiple-fault diagnosis [33,35] and extended to perform principal components analysis [33].

A.6 DIAGNOSTIC STRATEGIES BASED ON QUALITATIVE MODELS

A diverse variety of diagnostic strategies based on qualitative models of process behavior have been suggested [18,36–42], being extended to principal components analysis [40] and methodologies to create sign-directed graphs automatically [39–42]. Such qualitative differential equations, called confluences, lie at the center of a reasoning approach called qualitative physics. Correctly describing physical systems with confluences is a procedure known as envisionment [43,44]. Although the solutions generated by confluences are inherently ambiguous, Forbus [45] suggests that such descriptions more closely follow the reasoning employed by expert problem solvers and thus lead to reasoning strategies that are easier for nonexperts to follow.[2]

The inherent problems with using qualitative simulation to describe process systems were discussed by Kramer and Oyeleye [36] and Kuipers [46]. The most debilitating of these problems is that such modeling eliminates too much useful information, thereby causing spurious solutions to be generated. To eliminate spurious diagnoses, Kramer and Oyeleye [36] were first to suggest a method for combining confluences with sign-directed graphs. Order-of-magnitude reasoning strategies have also been suggested as a means to help reduce the inherent ambiguity of diagnostic methods based on qualitative reasoning [47–49]. A detailed review of qualitative diagnostic strategies as used in the CPI is given by Venkatasubramanian et al. [50].

A.7 DIAGNOSTIC STRATEGIES BASED ON QUANTITATIVE MODELS

Overcoming the diagnostic ambiguity inherent in all the strategies discussed previously requires additional diagnostic information. For process fault analysis, such information is usually available in the form of quantitative models derived from first principles describing the target process system's normal operating behavior. In general, this *deep knowledge* [51,52] of the process system not only improves the fault analyzer's ability to better discriminate between the various possible fault situations but typically, also improves the sensitivity at which the fault analyzer can diagnose each of those fault situations. The key feature common to most diagnostic strategies based on quantitative models is that they use redundant sources of information (i.e., *analytical redundancy*) to detect discrepancies. This redundancy typically arises from having both calculated and measured values of process variables available.

---

[2]This assertion was supported by actual experiments conducted with process experts during the original FALCON project. The results of these experiments are reported in Appendix B.

In general, the major drawback with basing diagnostic strategies on quantitative models lies in the fact that sufficiently accurate models are sometimes difficult, if not impossible, to derive. Such models also require greater computer resources to solve and analyze.

Many quantitative diagnostic strategies are based on estimation of state variables and unmeasured process parameters with Kalman filters [25,53–67]. These techniques use accurate models to estimate process variables and model parameters directly from measured process variables. These estimates are then compared via statistical tests to reference values obtained under normal operating conditions. Significant discrepancies in these comparisons indicate the presence of process faults.

Himmelblau [25] outlined some of the limitations of the early diagnostic methods based on Kalman filters. The limitations included that (1) these techniques performed poorly for process systems that could not be approximated with linear models; (2) they could not take into account the response of control systems; (3) to work properly, the filters typically required a high degree of tuning; and (4) the stability of the calculation could not be guaranteed in all situations. Park and Himmelblau [55] and King and Gilles [60] describe methods for extending these techniques to nonlinear models. Fathi et al. [64,65] extended these techniques to knowledge-based system approaches. The other advances in these techniques have been surveyed by Isermann [57], Himmelblau [58], Mah [62], Kramer and Mah [66], and Venkatasubramanian et al. [67]. However, it is doubtful that these techniques will ever be used to diagnose highly disruptive process faults or to monitor process operation during process startups, emergency process shutdowns, and so on [61].

Many other quantitative diagnostic strategies rely on the availability of redundant information. A method for detecting measurement errors and process leaks through redundancy classification was presented by Mah et al. [68]. Stephanopoulos and Romagnoli [69] also used these principles to develop a serial elimination algorithm for diagnosing measurement errors. Kretsovalis and Mah [70] have presented a more general algorithm for performing redundancy classification. Gertler and co-workers [71–73] have developed a method based on redundancy which ensures that the resulting fault diagnoses will always be correct. However, this method does not optimize the diagnostic sensitivity of the resulting fault analyzer for each possible fault situation, nor does it handle multiple-fault situations. Other strategies for incorporating quantitative models into process fault diagnosis have been presented by Lind [74], Wu [75], Petti et al. [76], Patton et al. [77], Frank [78,79], Chang et al. [80,81], Dunia et al. [82], and Gertler et al. [83,84]. Qin and Li [85] have extended these techniques by calculating EWMAs on errors in the models' residuals and signals with the underlying fault magnitudes being estimated directly from those residuals. Furthermore, formal theories of fault diagnosis based on first principles using analytical redundancy have been proposed by both Kramer [86] and Reiter [87].

A variety of diagnostic strategies centered on model-based reasoning also exist in other problem domains besides process fault diagnosis. One domain that has received a lot of attention is electronic circuit diagnosis. Both Davis and co-workers [88–91] and Genesereth [92,93] have developed algorithms for reasoning from circuit structure and component function to diagnose circuit malfunctions. This work has been extended by de Kleer and Williams [94] to handle multiple-fault situations.

A.8 ARTIFICIAL NEURAL NETWORK STRATEGIES

Good discussions of artificial neural networks and summaries of their application to process fault analysis are given by Himmelblau [95] and Venkatasubramanian et al. [96]. Briefly, neural networks are trained to identify various process faults by altering the connection weights within the neural network via recursive back-propagation calculations. They are advantageous for process fault analysis because an a priori model of the target process operation is not required. Watanabe et al. [97] demonstrated competent performance with a two-tiered neural network, the first level identifying the fault while the second estimates its magnitude.

The problem with neural networks is that adequate training data (e.g., data collected during a particular fault situation and at a variety of possible fault magnitudes or rates of occurrence) is typically unavailable for all the faults of interest. As discussed in Appendix B, actual fault data from DuPont's adipic acid process existed only for a mere handful of possible faults and at one or so specific fault magnitudes. Furthermore, even if fault data are available, neural network training is computationally intensive and highly dependent on the underlying organization of the number of nodes and number of node layers utilized. If trained properly, however, such networks are capable of extrapolating and interpolating current patterns in the process data to identify fault hypotheses within their domain.

A.9 KNOWLEDGE-BASED SYSTEM STRATEGIES

The knowledge-based system (also commonly referred to as the *expert system*) approach to problem solving is a paradigm that evolved originally within the field of computer science called artificial intelligence. In reality, this approach encompasses a vast collection of diverse problem-solving strategies. The common characteristic uniting all of these strategies is that when it comes to problem solving, each unequivocally stresses the dominance of domain-specific knowledge over the particular problem-solving strategy being used. This approach has recently found widespread support because it has been applied successfully to a variety of difficult problems that previously could be solved only by human experts.

Topics and issues being investigated within the field of artificial intelligence, including the subfield of knowledge-based systems, are discussed in many publications [98–110]. Other publications [111–134] discuss the general issues associated with developing and using knowledge-based systems. Unfavorable views of artificial intelligence in general and of the knowledge-based system approach in particular are also available in the published literature [135–140].

The biggest advantage of knowledge-based systems lies in their unlimited flexibility when it comes to combining domain knowledge obtained from diverse sources, such as the experiential heuristic knowledge commonly used by domain experts. Strategies proposed for combining the various sources of domain knowledge used in fault diagnosis can also be found in the literature [23,52,141–174]. Details on the performance of diagnostic knowledge base systems developed for actual applications are available in these publications [134,141,142,144–146,149,150,168].

Knowledge-based system approaches have also been suggested for automating other facets of process control and operation besides process fault analysis [175–183]. Specific strategies have been outlined for using expert systems in both the design [184] and operation [185] of process control systems and as general-purpose operator aids in process system operation [186]. Benefits of applying real-time knowledge-based systems in the pharmaceutical industry for this purpose have been reported by Eli Lilly & Co. [187–189]. The biggest advantages from their use of this technology has been that it frees up process experts to perform additional value-added activities besides such process operations while also reducing process variability and improving process yields. Appendix B describes a real-time knowledge-based system called FALCON. Its major motivation was to study various issues involved with building real-time online knowledge-based systems for actual process control applications, specifically those associated with automating process fault analysis.

A.10 METHODOLOGY CHOICE CONCLUSIONS

The common theme between all these various methodologies for automating process fault analysis is that they try to identify faults from patterns of diagnostic evidence (i.e., symptoms) that occur during those faults. All are equally valid means of doing this. However, even if implemented correctly, their various specific limitations regarding diagnostic resolution and diagnostic sensitivity hinder their usefulness in actual process applications. This has led our search for the most optimal diagnostic methodology, in both actual performance and cost-effectiveness, so that automating process fault analysis will become more commonplace in the chemical and nuclear processing industries. We contend that the MOME diagnostic strategy as codified in FALCONEER^{TM} IV is the best means of accomplishing this. This judgment is based on the extensive research and development that we have carried out while trying to automate process fault analysis in real-world applications. Evaluating and analyzing quantitative models of normal process operation simply provides the most unimpeachable source of process knowledge from which to analyze current process operations. This has been confirmed repeatedly by the competent performance of the resulting knowledge-based systems in actual process applications.

REFERENCES

1. Roland, H. E., and B. Moriarity, "Fault Tree Analysis" in *System Safety Engineering and Management*, Wiley, New York, 1983, pp. 215–271.

2. Hauptmanns, U., and J. Yllera, "Fault-Tree Evaluation by Monte Carlo Simulation," *Chemical Engineering*, January 1983, pp. 91–97.

3. Petersen, H. J. S., and P. Haastrup, "Fault-Tree Evaluation Shows Importance of Testing Instruments and Controls," *Chemical Engineering*, November 1984, pp. 85–87.

4. Lapp, S. A., and G. J. Powers, "Computer-Aided Synthesis of Fault Trees," *IEEE Transactions on Reliability*, Vol. R-26, April 1977, pp. 2–13.

5. Cummings, D. L., S. A. Lapp, and G. J. Powers, "Fault Tree Synthesis from a Directed Graph Model for a Power Distribution Network," *IEEE Transactions on Reliability*, Vol. R-32, No. 2, 1983, pp. 140–149.

6. Andow, P. K., "Difficulties in Fault-Tree Synthesis for Process Plant Diagnosis," *IEEE Transactions on Reliability*, Vol. R-29, No. 1, 1980, pp. 2–8.

7. Carpignano, A., and A. Poucet, "Computer-Assisted Fault Tree Construction: A Review of Methods and Concerns," *Reliability Engineering and System Safety*, Vol. 44, No. 3, 1994, pp. 256–278.

8. Srinivasan, R., and V. Venkatsubramanian, "Multi-perspective Models for Process Hazards Analysis of Large Scale Chemical Processes," *Computers and Chemical Engineering*, Vol. 22, Suppl. S, 1998, pp. S961–S964.

9. Meel, A., and W. D. Seider, "Plant Specific Dynamic Failure Assessment using Baysian Theory," *Chemical Engineering Science*, Vol. 61, No. 21, 2006, pp. 7036–7056.

10. Meel, A., and W. D. Seider, "Real Time Risk Analysis of Safety Systems," *Computers and Chemical Engineering*, Vol. 32, No. 4–5, 2008, pp. 827–840.

11. Kim, J., J. Kim, Y. Lee, et al., "Development of a New Automatic System for Fault Tree Analysis for Chemical Process Industries," *Korean Journal of Chemical Engineering*, Vol. 26, No. 6, 2009, pp. 1429–1440.

12. Martin-Solis, G. A., P. K. Andow, and F. P. Lees, "Fault Tree Synthesis for Design and Real Time Applications," *Transactions of the Institution of Chemical Engineers*, Vol. 60, 1982, pp.14–25.

13. Powers, G. J., "Probabilistic Risk and Reliability Assessment for Process Operations," in *Proceedings of the First International Conference on Foundations of Computer Aided Process Operations*, ed. by G. V. Reklaitis and H. D. Spriggs, Elsevier Science, New York, 1987, pp. 199–213.

14. Ulerich, N. B., and G. J. Powers, "Real-Time Hazard Aversion and Fault Diagnosis in Chemical Processes: A Model-Based Algorithm," Paper 83d, presented at the *AIChE Spring National Meeting*, Houston, TX, April 1987.

15. Yoon, E. S., and J. H. Han, "Process Failure Detection and Diagnosis Using the Tree Model," in *IFAC Kyoto Workshop on Fault Detection and Safety in Chemical Plants*, 1986, pp. 126–131.

16. Arueti, S., W. C. Gekler, M. Kazarians, and S. Kaplan, "Integrated Risk Assessment Program for Risk Management," Paper 83e, presented at the *AIChE Spring National Meeting*, Houston, TX, April 1987.

17. Arendt, J. S., and M. L. Casada, "Prisim: An Expert System for Process Risk Management," Paper 82e, presented at the *AIChE Spring National Meeting*, Houston, TX, April 1987.

18. Andow, P. K, and F. P. Lees, "Process Computer Alarm Analysis: Outline of a Method Based on List Processing," *Transactions Institution of Chemical Engineers,* Vol. 53, 1975, pp. 195–208.

19. Lees, F. P., "Computer Support for Diagnostic Tasks in the Process Industries," in *Human Detection and Diagnosis of System Failures*, ed. by J. Rasmussen and W. Rouse, Plenum Press, New York, 1981.

20. Bastl, W., and L. Felkel, "Disturbance Analysis Systems," in *Human Detection and Diagnosis of System Failures*, ed. by J. Rasmussen and W. B. Rouse, Plenum Press, New York, 1981, pp. 451–472.

21. Lees, F. P., "Process Computer Alarm and Disturbance Analysis: Review of the State of the Art," *Computers and Chemical Engineering*, Vol. 7, No. 6, 1983, pp. 669–694.

22. Lees, F. P., "Process Computer Alarm and Disturbance Analysis: Outline of Methods for Systematic Synthesis of the Fault Propagation Structure," *Computers and Chemical Engineering*, Vol. 8, No. 2, 1984, pp. 91–103.

23. Andow, P. K., "Fault Diagnosis Using Intelligent Knowledge Based Systems," *Chemical Engineering Design*, Vol. 63, November 1985, pp. 368–372.

24. Berenblut, B. J., and H. B. Whitehouse, "A Method for Monitoring Process Plant Based on a Decision Table Analysis," *Chemical Engineer*, March 1977, pp. 175–181.

25. Himmelblau, D. M., *Fault Detection and Diagnosis in Chemical and Petrochemical Processes*, Elsevier Scientific, Amsterdam, 1978.

26. Lihou, D. A., "Aiding Process Plant Operators in Fault Finding and Corrective Action," in *Human Detection and Diagnosis of System Failures*, ed. by J. Rasmussen and W. B. Rouse, Plenum Press, New York, 1981, pp. 501–522.

27. Iri, M., K. Aoki, E. O'Shima, and H. Matsushima, "An Algorithm for Diagnosis of System Failures in the Chemical Process," *Computers and Chemical Engineering*, Vol. 3, 1981, pp. 489–493.

28. Palowitch, B. L., and M. A. Kramer, "The Application of a Knowledge-Based Expert System to Chemical Plant Fault Diagnosis," presented at the *American Control Conference*, Boston, June 1985.

29. Shiozaki, J., H. Matsuyama, E. O'Shima, and M. Iri, "An Improved Algorithm for Diagnosis of System Failures in the Chemical Process," *Computers and Chemical Engineering*, Vol. 9, No. 3, 1985, pp. 285–293.

30. Kramer, M. A., and B. L. Palowitch, "A Rule-Based Approach to Fault Diagnosis Using the Signed Directed Graph," *AIChE Journal*, Vol. 33, 1987, pp. 1067–1088.

31. Shibata, B., Y. Tsuge, H. Matsuyama, and E. O'Shima, "A Fault Diagnosis System for the Continuous Process with Frequent Load-Fluctuations," in *IFAC Kyoto Workshop on Fault Detection and Safety in Chemical Plants*, 1986, pp. 132–136.

32. Wang, X. Z., S. A. Yang, E. Veloso, et al., "Qualitative Process Modeling: A Fuzzy Sign Directed Graph Method," *Computers and Chemical Engineering*, Vol. 19, Suppl. S, 1995, pp. S735–S740.

33. Vedam, H., and V. Venkatsubramanian, "PCA-SDG Based Process Monitoring and Fault Diagnosis," *Control Engineering Practice*, Vol. 7, No. 7, 1999, pp. 903–917.

34. Maurya, M. R., R. Rengaswamy, and V. Venkatsubramanian, "A Systematic Framework for the Development and Analysis of Signed Diagraphs for Chemical Processes: Part 1. Algorithms and Analysis," *Industrial and Engineering Chemistry Research*, Vol. 42, No. 20, 2003, pp. 4789–4810.

35. Zhang, Z. Q., C. G. Wu, B. K. Zhang, et al., "SDG Multiple Fault Diagnosis by Real-Time Inverse Reference," *Reliability Engineering and System Safety*, Vol. 87, No. 2, 2005, pp. 173–189.

36. Kramer, M. A., and O. O. Oyeleye, "Qualitative Simulation of Chemical Process Plants," presented at the *IFAC Tenth World Congress*, Munich, Germany, 1987.

37. Rich, S. H., and V. Venkatasubramanian, "Model-Based Reasoning in Diagnostic Expert Systems for Chemical Process Plants," *Computers and Chemical Engineering*, Vol. 11, No. 2, 1987, pp. 111–122.

38. Bourseau, P., K. Bousson, P. Daugue, et al., "Qualitative Reasoning: A Survey of Techniques and Applications," *AI Communications*, Vol. 8, Nos. 3–4, 1995, pp. 119–192.

39. Maurya, M. R., R. Rengaswamy, and V. Venkatsubramanian, "A Sign Directed Graph and Qualitative Trend Analysis Based Framework for Incipient Fault Diagnosis," *Chemical Engineering Research and Design*, Vol. 85, No. A10, 2007, pp. 1407–1422.

40. Maurya, M. R., R. Rengaswamy, and V. Venkatsubramanian, "Fault Diagnosis by Qualitative Trend Analysis of the Principal Components," *Chemical Engineering Research and Design*, Vol. 83, No. A9, 2005, pp. 1122–1132.

41. Gao, D., C. Wu, B. Zhang, et al., "Sign Directed Graph and Qualitative Trend Analysis Based Fault Diagnosis in the Chemical Industry," *Chinese Journal of Chemical Engineering*, Vol. 18, No. 2, 2010, pp. 265–276.

42. Zhang, W., C. Wu, and C. Wang, "Qualitative Algebra and Graph Theory Methods for Dynamic Trend Analysis of Continuous Systems," *Chinese Journal of Chemical Engineering*, Vol. 19, No. 2, 2011, pp. 308–315.

43. Kuipers, B., "Getting the Envisionment Right," in *Proceedings of the National Conference on Artificial Intelligence*, Washington, DC, 1983, pp. 209–212.

44. de Kleer, J., and J. S. Brown, "Foundations of Envisioning," in *Proceedings of the National Conference on Artificial Intelligence*, Washington, DC, 1983, pp. 434–437.

45. Forbus, K. D., "Intelligent Computer-Aided Engineering," presented at the *AAAI Workshop on Artificial Intelligence in Process Engineering*, Columbia University, New York, March 1987.

46. Kuipers, B., "The Limits of Qualitative Simulation," in *Proceedings of the Ninth International Joint Conference on Artificial Intelligence*, Los Angeles, Vol. 1, ed. by A. Joshi, Morgan Kaufmann Publishers, Los Altos, CA, 1985, pp. 128–136.

47. Raiman, O., "Order of Magnitude Reasoning," in *Proceedings of the Fifth National Conference on Artificial Intelligence*, Philadelphia, Vol. 1, Morgan Kaufmann Publishers, Los Altos, CA, 1986, pp. 100–104.

48. Dague, P., O. Raiman, and P. Deves, "Troubleshooting: When Modeling Is the Trouble," in *Proceedings of the Sixth National Conference on Artificial Intelligence*, Seattle, WA, Vol. 1, Morgan Kaufmann Publishers, Los Altos, CA, 1987, pp. 600–605.

49. Mavrovouniotis, M. L., and G. Stephanopoulos, "Reasoning with Orders of Magnitude and Approximate Relations," in *Proceedings of the Sixth National Conference on Artificial Intelligence*, Seattle, WA, Vol. 1, Morgan Kaufmann Publishers, Los Altos, CA, 1987, pp. 626–630.

50. Venkatasubramanian, V., R. Rengaswamy, and S. N. Kavuri, "A Review of Process Fault Detection and Diagnosis: Part II. Qualititive Models and Search Strategies," *Computers and Chemical Engineering*, Vol. 27, 2003, pp. 293–311.

51. Hart, P. E., "Directions for AI in the Eighties," *SIGART Newsletter*, January 1982, pp. 11–16.

52. Michie, D., "High-Road and Low-Road Programs," *AI Magazine*, Winter 1981–1982, pp. 21–22.

53. Stanley, G. M., and R. S. H. Mah, "Estimation of Flows and Temperatures in Process Networks," *AIChE Journal*, Vol. 23, No. 5, September 1977, pp. 642–650.

54. Mah, R. S. H., and A. C. Tamhane, "Detection of Gross Errors in Process Data," *AIChE Journal*, Vol. 28, No. 5, September, 1982, pp. 828–830.

55. Park, S., and D. M. Himmelblau, "Fault Detection and Diagnosis via Parameter Estimation in Lumped Dynamic Systems," *Industrial and Engineering Chemistry Process Design*, Vol. 22, No. 3, 1983, pp. 482–487.

56. Tylee, J. L., "On-Line Failure Detection in Nuclear Power Plant Instrumentation," *IEEE Transactions on Automatic Control*, Vol. AC-28, No. 3, March 1983, pp. 406–415.

57. Isermann, R., "Process Fault Detection Based on Modeling and Estimation Methods: A Survey," *Automatia*, Vol. 30, 1984, pp. 387–404.

58. Himmelblau, D. M., Keynote Lecture: "Fault Detection and Diagnosis: Today and Tomorrow," in *IFAC Kyoto Workshop on Fault Detection and Safety in Chemical Plants*, Kyoto, Japan, 1986, pp. 95–105.

59. Halme, A., and J. Selkainaho, "An Adaptive Filtering Based Method to Detect Sensor/Actuator Faults," in *IFAC Kyoto Workshop on Fault Detection and Safety in Chemical Plants*, Kyoto, Japan, 1986, pp. 158–162.

60. King, R., and E. D. Gilles, "Early Detection of Hazardous States in Chemical Reactors," in *IFAC Kyoto Workshop on Fault Detection and Safety in Chemical Plants*, Kyoto, Japan, 1986, pp. 137–143.

61. Hengy, D., and P. M. Frank, "Component Failure Detection via Nonlinear State Observers," in *IFAC Kyoto Workshop on Fault Detection and Safety in Chemical Plants*, Kyoto, Japan, 1986, pp. 153–157.

62. Mah, R. S. H., "Data Screening," in *Proceedings of the First International Conference on Foundations of Computer Aided Process Operations*, Park City, UT, ed. by G. V. Reklaitis and H. D. Spriggs, Elsevier Science, New York, July 1987, pp. 67–94.

63. Narasimhan, S., and R. S. H. Mah, "Gross Error Identification in Dynamic Processes Using Likelihood Ratios," *AIChE Journal*, Vol. 33, 1987, pp. 1514–1521.

64. Fathi, Z., W. F. Ramirez, and J. Korbicz, "Analytical and Knowledge-Based Redundancy for Fault Diagnosis in Process Plants," *AIChE Journal*, Vol. 39, No. 1, 1993, pp. 42–56.

65. Fathi, Z., W. F. Ramirez, A. P. Tavares, et al., "Symbolic Reasoning and Quantitative Analysis for Fault Detection and Isolation in Process Plants," *Engineering Applications of Artificial Intelligence*, Vol. 6, No. 3, 1993, pp. 203–218.

66. Kramer, M. A., and R. S. H. Mah, "Model-Based Monitoring," in *Foundations of Computer-Aided Process Operations II*, ed. by D. W. T. Rippin, J. C. Hale, and J. F. Davis, CACHE, Inc., Austin, TX, 1994, pp. 45–68.

67. Venkatasubramanian, V., R. Rengaswamy, K. Yin, and S. N. Kavuri, "A Review of Process Fault Detection and Diagnosis: Part I. Quantatitive Model-Based Methods," *Computers and Chemical Engineering*, Vol. 27, 2003, pp. 313–326.

68. Mah, R. S., G. M. Stanley, and D. M. Downing, "Reconciliation and Rectification of Process Flow and Inventory Data," *Industrial and Engineering Chemistry Process Design*, Vol. 15, No. 1, 1976, pp. 175–183.

69. Stephanopoulos, G., and J. A. Romagnoli, "A General Approach to Classify Operational Parameters and Rectify Measurement Errors for Complex Chemical Processes," in *Computer Applications to Chemical Engineering*, American Chemical Society, Washington, DC, 1980, pp. 153–174.

70. Kretsovalis, A., and R. S. H. Mah, "Observability and Redundancy Classification in Multicomponent Process Networks," *AIChE Journal*, Vol. 33, No. 1, January 1987, pp. 70–82.

71. Gertler, J., "Fault Detection and Isolation in Complex Technical Systems – A New Model Error Approach," Conference on Information Sciences and Systems, Johns Hopkins University, Baltimore, MD, 1985.

72. Gertler, J., D. Singer, and A. Sundar, "A Robustified Linear Fault Isolation Technique for Complex Dynamic Systems," in *IFAC Symposium on Digital Computer Applications to Process Control*, Vienna, Austria, 1985, pp. 493–498.

73. Gertler, J., and D. Singer, "Augmented Models for Statistical Fault Isolation in Complex Dynamic Systems," in *American Control Conference*, Boston, 1985, pp. 317–322.

74. Lind, M., "The Use of Flow Models for Automated Plant Diagnosis," in *Human Detection and Diagnosis of System Failures*, ed. by J. Rasmussen and W. B. Rouse, Plenum Press, New York, 1981, pp. 411–432.

75. Wu, R. S. H., "Dynamic Thermal Analyzer for Monitoring Batch Processes," *Chemical Engineering Progress*, September 1985, pp. 57–61.

76. Petti, T. F., J. Klein, and P. S. Dhurjati, "Diagnostic Model Processor: Using Deep Knowledge for Process Fault Diagnosis," *AIChE Journal*, Vol. 36, No. 4, April 1990, pp. 565–575.

77. Patton, R., P. M. Frank, and R. Clark, *Fault Diagnosis in Dynamic Systems: Theory and Application,* Prentice Hall, Englewood Cliffs, NJ, 1989.

78. Frank, P. M., "Fault Diagnosis in Dynamic Systems Using Analytical and Knowledge-Based Redundancy: A Survey and Some New Results," *Automatica*, Vol. 26, No. 3, 1990, pp. 459–474.

79. Frank, P. M., "Robust Model-Based Fault Detection in Dynamic Systems," in *Proceedings of the On-Line Fault Detection and Supervision in the Chemical Process Industries,* IFAC Symposium, ed. by P. S. Dhurjati, Newark, DE, 1992, pp. 1–13.

80. Chang, I. C., C. C. Yu, and C. T. Liou, "Model-Based Approach for Fault Analysis: 1. Principles of Deep Model Algorithm," *Industrial and Engineering Chemistry Research*, Vol. 33, 1994, pp. 1542–1555.

81. Chang, I. C., C. C. Yu, and C. T. Liou, "Model-Based Approach for Fault Analysis: Part 2. Extension to Interval Systems," *Industrial and Engineering Chemistry Research*, Vol. 34, 1994, pp. 828–844.

82. Dunia, R., S. J. Qin, T. F. Edgar, and T. J. McAvoy, "Identification of Faulty Sensors Using Principal Component Analysis," *AIChE Journal*, Vol. 42, No. 10, October 1996, pp. 2797–2812.

83. Gertler, J., "Structured Residuals for Fault Isolation, Disturbance Decoupling and Modelling Error Robustness," in *Proceedings of the On-Line Fault Detection and Supervision in the Chemical Process Industries*, IFAC Symposium, ed. by P. S. Dhurjati, Newark, DE, 1992, pp. 111–119.

84. Gertler, J., W. Li, Y. Huang, and T. J. McAvoy, "Isolation Enhanced Principal Component Analysis," *AIChE Journal*, Vol. 45, No. 2, February 1999, pp. 323–334.

85. Qin, S. J., and W. H. Li, "Detection and Identification of Faulty Sensors in Dynamic Processes," *AIChE Journal*, Vol. 47, No. 7, 2001, pp. 1581–1593.

86. Kramer, M. A., "Malfunction Diagnosis Using Quantitative Models and Non-Boolean Reasoning in Expert Systems," *AIChE Journal*, Vol. 33, 1987, pp. 130–147.

87. Reiter, R., "A Theory of Diagnosis from First Principles," *Artificial Intelligence*, Vol. 32, 1987, pp. 57–95.

88. Davis, R., H. Shrobe, W. Hamscher, K. Wieckert, K., M. Shirley, and S. Polit, "Diagnosis Based on Description of Structure and Function," in *Proceedings of the National Conference on Artificial Intelligence*, Pittsburgh, PA, 1982, pp. 137–142.

89. Davis, R., "Reasoning from First Principles in Electronic Troubleshooting," *International Journal of Man–Machine Studies*, Vol. 19, No. 5, November 1983, pp. 403–423.

90. Hamscher, W., and R. Davis, "Diagnosing Circuits with State: An Inherently Underconstrained Problem," in *Proceedings of the National Conference on Artificial Intelligence*, Austin, TX, 1984, pp. 142–147.

91. Davis, R., "Diagnostic Reasoning Based on Structure and Behavior," *Artificial Intelligence*, Vol. 24, 1984, pp. 347–410.

92. Genesereth, M. R., "Diagnosis Using Hierarchical Design Models," in *Proceedings of the National Conference on Artificial Intelligence*, Pittsburgh, PA, 1982, pp. 278–283.

93. Genesereth, M. R., "The Use of Design Descriptions in Automated Diagnosis," *Artificial Intelligence*, Vol. 24, 1984, pp. 411–436.

94. de Kleer, J., and B. C. Williams, "Diagnosing Multiple Faults," *Artificial Intelligence*, Vol. 32, 1987, pp. 97–130.

95. Himmelblau, D. M., "Use of Artificial Neural Networks to Monitor Faults and for Troubleshooting in the Process Industries," in *Proceedings of the On-Line Fault Detection and Supervision in the Chemical Process Industries*, IFAC Symposium, ed. by P. S. Dhurjati, Newark, DE, 1992, pp. 144–149.

96. Venkatasubramanian, V., R. Rengaswamy, S. N. Kavuri, and K. Yin, "A Review of Process Fault Detection and Diagnosis: Part III. Process History Based Methods," *Computers and Chemical Engineering*, Vol. 27, 2003, pp. 327–346.

97. Watanabe, K., I. Matsuura, M. Abe, M. Kubota, and D. M. Himmelblau, "Incipient Fault Diagnosis of Chemical Processes via Artificial Neural Networks," *AIChE Journal*, Vol. 35, No. 11, November 1989, pp. 1803–1812.

98. Bobrow, D. G., and P. J. Hayes, eds., "Artificial Intelligence: Where Are We?" *Artificial Intelligence*, Vol. 25, 1985, pp. 375–415.

99. Nilsson, N. J., "Artificial Intelligence: Engineering, Science, or Slogan?" *AI Magazine*, Vol. 2, No. 4, Winter 1982, pp. 2–9.

100. Schank, R. C., "The Current State of AI: One Man's Opinion," *AI Magazine*, Vol. 4, No. 1, Winter–Spring 1983, pp. 3–8.

101. Feigenbaum, E. A., "The Art of Artificial Intelligence: Themes and Case Studies of Knowledge Engineering," in *Fifth International Joint Conference on Artificial Intelligence*, Cambridge, MA, August 1977, pp. 1014–1029.

102. Lenat, D. B., "The Nature of Heuristics," *Artificial Intelligence*, Vol. 19, 1982, pp. 189–249.

103. Newell, A., "The Knowledge Level," *Artificial Intelligence*, Vol. 18, 1982, pp. 87–127.

104. Brachman, R., and H. J. Levesque, "Competence in Knowledge Representation," in *Proceedings of the National Conference on Artificial Intelligence*, Pittsburgh, PA, August 1982, pp. 189–192.

105. Clancey, W. J., "Classification Problem Solving," in *Proceedings of the National Conference on Artificial Intelligence*, Austin, TX, August 1984, pp. 49–55.

106. Clancey, W. J., "Heuristic Classification," *Artificial Intelligence*, Vol. 27, 1985, pp. 289–350.

107. Charniak, E., and D. McDermott, *Artificial Intelligence*, Addison-Wesley Reading, MA, 1985.

108. Rich, E., *Artificial Intelligence*, McGraw-Hill, New York, 1983.

109. Brachman, R. J., and H. J. Levesque, eds., *Readings in Knowledge Representation*, Morgan Kaufmann Publishers, Los Altos, CA, 1985.

110. Genesereth, M. R., and N. J. Nilsson, *Logical Foundations of Artificial Intelligence*, Morgan Kaufmann Publishers, Los Altos, CA, 1987.

111. Stefik, M., J. Aikins, R. Balzer, J. Benoit, L. Birnbaum, F. Hayes-Roth, and E. Sacerdoti, "The Organization of Expert Systems, A Tutorial," *Artificial Intelligence*, Vol. 18, 1982, pp. 135–173.

112. Clancey, W., "The Epistemology of a Rule-Based Expert System: A Framework for Explanation," *Artificial Intelligence*, Vol. 20, 1983, pp. 215–251.

113. Davis, R., "Meta-rules: Reasoning About Control," *Artificial Intelligence*, Vol. 15, 1980, pp. 179–222.

114. Davis, R., "Content Reference: Reasoning About Rules," *Artificial Intelligence*, Vol. 15, 1980, pp. 223–239.

115. Gevarter, W. B., "The Nature and Evaluation of Commercial Expert System Building Tools," *Computer*, May 1987, pp. 24–41.

116. Weiss, S. M., and C. A. Kulikowski, *A Practical Guide to Designing Expert Systems*, Rowman & Allanheld, Totowa, NJ, 1984.

117. Buchanan, B., and E. H. Shortliffe, eds., *Rule-Based Expert Systems*, Addison-Wesley, Reading, MA, 1984.

118. Freiling, M., J. Alexander, S. Messick, S. Rehfuss, and S. Shulman, S., "Starting a Knowledge Engineering Project: A Step by Step Approach," *AI Magazine*, Vol. 6, No. 3, Fall 1985, pp. 150–164.

119. Prerau, D. S., "Knowledge Acquisition in the Development of a Large Expert System," *AI Magazine*, Vol. 8, No. 2, Summer 1987, pp. 43–51.

120. Hoffman, R., "The Problem of Extracting the Knowledge of Experts from the Perspective of Experimental Psychology," *AI Magazine*, Vol. 8, No. 2, Summer 1987, pp. 53–67.

121. Sridharan, N. S., "Evolving Systems of Knowledge," *AI Magazine*, Vol. 6, No. 3, Fall 1985, pp. 108–120.

122. Hink, R. F., and D. L. Woods, "How Humans Process Uncertain Knowledge: An Introduction for Knowledge Engineers," *AI Magazine*, Vol. 8, No. 3, Fall 1987, pp. 41–53.

123. Hayes-Roth, F., D. A. Waterman, and D. B. Lenat, eds., *Building Expert Systems*, Addison-Wesley, Reading, MA, 1983.

124. Jackson, P., *Introduction to Expert Systems*, Addison-Wesley, Workingham, England, 1986.

125. Sell, P. S., *Expert Systems: A Practical Introduction*, Wiley, New York, 1985.

126. Davis, R., "Expert Systems: Where Are We? And Where Do We Go from Here?" *AI Magazine*, Vol. 3, No. 1, Spring 1982, pp. 3–22.

127. Davis, R., "Amplifying Expertise with Expert Systems," in *The AI Business,* ed. By P. H. Winston and K. A. Prendergest, MIT Press, Cambridge, MA, 1984, pp. 17–40.

128. Stefik, M., J. Aikins, R. Balzer, J. Benoit, L. Birnbaum, F. Hayes-Roth, and E. Sacerdoti, "The Organization of Expert Systems, A Tutorial," *Artificial Intelligence*, Vol. 18, 1982, pp. 135–173.

129. Basden, A., "On the Application of Expert Systems," in *Developments in Expert Systems*, Academic Press, London, 1984, pp. 59–75.

130. Chandrasekaren, B., "Towards a Taxonomy of Problem Solving Types," *AI Magazine*, Winter–Spring 1983, pp. 9–17.

131. Kline, P. J., and S. B. Dolins, "Choosing Architectures for Expert Systems," *Texas Instruments Inc. CCSC Technical Report, 85-01-001*, Dallas, TX, October 1985.

132. Kline, P., and S. Dolins, "Problem Features That Influence the Design of Expert Systems," in *Proceedings of the Fifth National Conference on Artificial Intelligence*, Philadelphia, Vol. 2, Morgan Kaufmann Publishers, Los Altos, CA, 1986, pp. 956–962.

133. Gevarter, W. B., "The Nature and Evaluation of Commercial Expert System Building Tools," *Computer*, May 1987, pp. 24–41.

134. Laffey, T. J., P. A. Cox, J. L. Schmidt, S. M. Kao, and J. Y. Read, "Real-Time Knowledge Based Systems," *AI Magazine*, Spring 1988, pp. 27–45.

135. Dreyfus, H. J., "From Micro-Worlds to Knowledge Representation: AI at an Impasse," in *Mind Design*, ed. by J. Haugeland, MIT Press, Cambridge, MA, 1981, pp. 161–204.

136. Dreyfus, H., and S. Dreyfus, "Why Computers May Never Think Like People," *Technology Review*, January 1986, pp. 42–61.

137. Martins, G. M., "The Overselling of Expert Systems," *Datamation*, November 1984, pp. 76–79.

138. Martins, G. M., "Smoke and Mirrors?" *Unix Review*, August 1987, p. 35.

139. Miline, R. W., "A Few Problems with Expert Systems," in *Proceedings of the IEEE Computer Society Conference*, McLean, VA, October 1985, pp. 66–67.

140. Simon, R., "The Morning After," *Forbes*, October 1987, pp. 165–168.

141. Gallanti, M., G. Giovanni, L. Spampinato, and A. Stefanini, "Representing Procedural Knowledge in Expert Systems: An Application to Process Control," in *Proceedings of the Ninth International Joint Conference on Artificial Intelligence*, Los Angeles, Vol. 1, ed. by A. Joshi, Morgan Kaufmann Publishers, Los Altos, CA, August 1985, pp. 345–352.

142. Fox, M. S., S. Lowenfeld, and P. Kleinosky, "Techniques for Sensor-Based Diagnosis," in *Proceedings on the International Joint Conference on Artificial Intelligence*, Karlsruhe, Germany, 1983, pp. 158–163.

143. Davis, R., "Diagnosis via Causal Reasoning: Paths of Interaction and the Locality Principle," in *Proceedings of the National Conference on Artificial Intelligence*, Washington, DC, 1983, pp. 88–94.

144. Pipitone, F., "The FIS Electronics Troubleshooting System," *Computer*, July 1986, pp. 68–76.

145. Underwood, W. E., "A CSA Model-Based Nuclear Power Plant Consultant," in *Proceedings of the National Conference on Artificial Intelligence*, Pittsburgh, PA, 1982, pp. 302–305.

146. Nelson, W. R., "Reactor: An Expert System for Diagnosis and Treatment of Nuclear Reactor Accidents," in *Proceedings of the National Conference on Artificial Intelligence*, Pittsburgh, PA, 1982, pp. 296–301.

147. Talukdar, S. N., E. Cardozo, and L. V. Leao, "Toast: The Power System Operator's Assistant," *Computer*, July 1986, pp. 53–60.

148. Yung-Choa Pan, J., "Qualitative Reasoning with Deep-Level Mechanism Models for Diagnoses of Mechanism Failures," *IEEE Spectrum*, 1984, pp. 295–301.

149. Yung-Choa Pan, J., and J. M. Tenenbaum, "PIES: An Engineer's Do-It-Yourself Knowledge System for Interpretation of Parametric Test Data," in *Proceedings of the Fifth National Conference on Artificial Intelligence*, Philadelphia, Vol. 2, Morgan Kaufmann Publishers, Los Altos, CA, 1986, pp. 836–843.

150. Yung-Choa Pan, J., and J. M. Tenenbaum, "Pies: An Engineer's Do-It-Yourself Knowledge System for Interpretation of Parametric Test Data," *AI Magazine*, Fall 1986, pp. 62–69.

151. Kramer, M. A., and B. L. Palowitch, "Expert System and Knowledge-Based Approaches to Process Malfunction Diagnosis," Paper 70b, presented at the *AIChE National Meeting*, Chicago, November 1985.

152. Venkatasubramanian, V., and S. H. Rich, "Integrating Heuristic and Deep-Level Knowledge in Expert Systems for Process Fault Diagnosis," presented at the *AAAI Workshop on Artificial Intelligence in Process Engineering*, Columbia University, New York, March 1987.

153. Chandrasekaran, B., and S. Mittal, "Deep Versus Compiled Knowledge Approaches to Diagnostic Problem-Solving," in *Proceedings of the National Conference on Artificial Intelligence*, Pittsburgh, PA, 1982, pp. 349–354.

154. Chandrasekaran, B., and S. Mittal, "Deep Versus Compiled Knowledge Approaches to Diagnostic Problem-Solving," in *Developments in Expert Systems*, ed. by M. J. Combs, Academic Press, London, 1984, pp. 23–34.

155. Bylander, T., and S. Mittal, "CSRL: A Language for Classificatory Problem Solving and Uncertainty Handling," *AI Magazine*, August 1986, pp. 66–77.

156. Miline, R. W., and B. Chandrasekaren, "Fault Diagnosis and Expert Systems," in *Sixth International Workshop on Expert Systems and Their Applications*, Avignon, France, April 1986, pp. 603–612.

157. Shum, S. K., J. F. Davis, W. F. Punch, and B. Chandrasedaran, "An Expert System Approach to Malfunction Diagnosis in Chemical Plants," *Computers and Chemical Engineering*, Vol. 12, No. 1, 1988, pp. 27–36.

158. Shum, S. K., J. F. Davis, W. F. Punch, and B. Chandrasedaran, "A Task-Oriented Approach to Malfunction Diagnosis in Complex Processing Plants," presented at the NSF-AAAI Workshop on AI in Process Engineering, Columbia University, New York, March 1987.

159. Davis, J. F., "A Task-Oriented Framework for Diagnostic and Design Expert Systems," in *Proceedings of the First International Conference on Foundations of Computer Aided Process Operations*, ed. by G. V. Reklaitis and H. D. Spriggs, Elsevier Science, New York, 1987, pp. 695–700.

160. Ramesh, T. S., S. K. Shum, and J. F. Davis, "A Structured Framework for Efficient Problem Solving in Diagnostic Expert Systems," *Computers and Chemical, Engineering*, Vol. 9–10, No. 12, 1988, pp. 891–902.

161. Thompson, W. B., P. E. Johnson, and J. B. Moen, "Recognition-Based Diagnostic Reasoning," in *Proceedings of the National Conference on Artificial Intelligence*, Washington, DC, 1983, pp. 236–238.

162. Reggia, J. A., D. S. Nau, and P. Y. Wang, "Diagnostic Expert Systems Based on a Set Covering Model," in *Developments in Expert Systems*, ed. by M. J. Combs, Academic Press, London, 1984, pp. 35–58.

163. Maletz, M. C., "An Architecture for Consideration of Multiple Faults," Second Conference on Artificial Intelligence Applications, in *Proceedings of the IEEE Computer Society Conference*, Miami Beach, FL, December 1985, pp. 60–67.

164. Lund, J. T., "Multiple Cause Identification in Diagnostic Problem Solving," *Technical Report 86–11*, Department of Computer and Information Sciences, University of Delaware, Newark, DE, 1986.

165. Hudlicka, E., and V. R. Lesser, "Meta-level Control Through Fault Detection and Diagnosis," in *Proceedings of the National Conference on Artificial Intelligence*, Austin, TX, August 1984, pp. 153–161.

166. Koton, P. A., "Empirical and Model-Based Reasoning in Expert Systems," in *Proceedings of the Ninth International Joint Conference on Artificial Intelligence*, Los Angeles, Vol. 1, ed. by A. Joshi, Morgan Kaufmann Publishers, Los Altos, CA, 1985, pp. 297–299.

167. Sticklen, J., B. Chandrasekaran, and J. R. Josephson, "Control Issues in Classificatory Diagnosis," in *Proceedings of the Ninth International Joint Conference on Artificial Intelligence*, Los Angeles, Vol. 1, ed. by A. Joshi, Morgan Kaufmann Publishers, Los Altos, CA, 1985, pp. 300–306.

168. Scarl, E. A., J. R. Jamieson, and C. L. Delaune, "A Fault Detection and Isolation Method Applied to Liquid Oxygen Loading for the Space Shuttle," in *Proceedings of the Ninth International Joint Conference on Artificial Intelligence*, Los Angeles, Vol. 1, ed. by A. Joshi, Morgan Kaufmann Publishers, Los Altos, CA, 1985, pp. 414–416.

169. Fink, P. K., "Control and Integration of Diverse Knowledge in a Diagnostic Expert System," in *Proceedings of the Ninth International Joint Conference on Artificial Intelligence*, Los Angeles, Vol. 1, ed. by A. Joshi, Morgan Kaufmann Publishers, Los Altos, CA, 1985, pp. 426–431.

170. Finch, F. E., and M. A. Kramer, "Narrowing Diagnostic Focus by Control System Decomposition," Paper 82b, presented at the *AIChE Spring National Meeting*, Houston, TX, 1987.

171. Xiang, Z., and S. N. Srihari, "A Strategy for Diagnosis Based on Empirical and Model Knowledge," in *Sixth International Workshop on Expert Systems and Their Applications*, Avignon, France, April 1986, pp. 835–848.

172. Kramer, M. A., "Integration of Heuristic and Model-Based Inference in Chemical Process Fault Diagnosis," presented at the *IFAC Workshop on Fault Detection and Safety in Chemical Plants*, Kyoto, Japan, September 1986.

173. Kramer, M. A., "Expert Systems for Process Fault Diagnosis: A General Framework," in *Proceedings of the First International Conference on Foundations of Computer Aided Process Operations*, ed. by G. V. Reklaitis and H. D. Spriggs, Elsevier Science, New York, 1987, pp. 557–589.

174. Chandrasekaran, B., and W. F. Punch, "Data Validation During Diagnosis: A Step Beyond Traditional Sensor Validation," in *Proceedings of the Sixth National Conference on Artificial Intelligence*, Seattle, WA, Vol. 1, Morgan Kaufmann Publishers, Los Altos, CA, 1987, pp. 778–782.

175. O'Shima, E., "Computer Aided Plant Operation," *Computers and Chemical Engineering*, Vol. 7, No. 4, 1983, pp. 311–329.

176. Moore, R. L., L. B. Hawkinson, C. G. Knickerbocker, and L. M. Churchman, "A Real-Time Expert System for Process Control," in *Proceedings of the First Conference on Artificial Intelligence* (IEEE), Denver, CO, 1984, pp. 528–540.

177. Stephanopoulos, G., "Expert Systems in Process Control," in *Proceedings of the Third International Conference on Chemical Process Control*, Asilomar, CA (A Cache Publication), Elsevier, New York, 1986, pp. 803–806.

178. Faught, W. S., "Applications of AI in Engineering," *Computer*, July 1986, pp. 17–27.

179. Taylor, J. H., "Expert Systems for Computer-Aided Control Engineering," in *Proceedings of the Third International Conference on Chemical Process Control*, Asilomac, CA (A CACHE Publication), Elsevier, New York, 1986, pp. 807–838.

180. Moore, R. L., and M. A. Kramer, "Expert Systems in On-Line Process Control," *Proceedings of the Third International Conference on Chemical Process Control*, Asilomar, CA (A CACHE Publication), Elsevier, New York, 1986, pp. 839–867.

181. Andow, P. K., "Goal-Based Control Systems," in *IFAC Kyoto Workshop on Fault Detection and Safety in Chemical Plants*, 1986, pp. 106–110.

182. Moore, R. L., B. Hawkinson, M. Levin, A. G. Hofmann, B. L. Matthews, and M. H. David, "Expert Systems Methodology for Real-Time Process Control," in *Proceedings of the 10th World Congress of Automatic Control*, IFAC, Munich, Germany, July 1987, Vol. 6, pp. 274–281.

183. Stephanopoulos, G., "The Scope of Artificial Intelligence in Plant-wide Operations," in *Proceedings of the First International Conference on Foundations of Computer Aided Process Operations*, ed. by G. V. Reklaitis and H. D. Spriggs Elsevier Science, New York, 1987, pp. 505–556.

184. Niida, K., and T. Umeda, "Process Control System Synthesis by an Expert System," in *Proceedings of the Third International Conference on Chemical Process Control*, Asilomac, CA (A CACHE Publication), Elsevier, New York, 1986, pp. 839–868.

185. Astrom, K. J., J. J. Anton, and K. E. Arzen, "Expert Control," *Automatia*, Vol. 22, No. 3, 1986, pp. 276–293.

186. Kaemmerer, W. F., and J. R. Allard, "An Automated Reasoning Technique for Providing Moment-by-Moment Advice Concerning the Operation of a Process," in *Proceedings of the Sixth National Conference on Artificial Intelligence*, Seattle, WA, Vol. 1, Morgan Kaufmann Publishers, Los Altos, CA, 1987, pp. 809–813.

187. Alford, J. S., C. Cairney, R. Higgs, M. Honsowetz, V. Huynh, A. Jines, D. Keates, and C. Skelton, "Real Rewards from Artificial Intelligence," *InTech*, April 1999, pp. 52–55.

188. Alford, J. S., C. Cairney, R. Higgs, M. Honsowetz, V. Huynh, A. Jines, D. Keates, and C. Skelton, "On-Line Expert System Applications Use in Fermentation Plants," *InTech*, July 1999, pp. 50–54.

189. Ranjan, A., J. Glassey, G. Montague, and P. Mohan, "From Process Experts to a Real-Time Knowledge-Based System," *Expert Systems*, Vol. 19, No. 2, 2002, pp. 69–79.

B

THE FALCON PROJECT

B.1 INTRODUCTION

The **FALCON** (fault analysis consultant) project was a joint venture of the University of Delaware, DuPont, and the Foxboro Company. Formally initiated in January 1984, its main objective was to develop a knowledge-based system capable of performing continuous real-time process fault analysis online in a commercial scale adipic acid plant operated by DuPont in Victoria, Texas. Officially commissioned at the plant on January 20, 1988, process operators used it online to analyze abnormal operating conditions in an adipic acid process. However, due to problems with the transmission of plant data to the FALCON system, it was decommissioned on April 15, 1988.

The major motivation for the FALCON project was to identify general issues involved with developing knowledge-based systems for actual process control applications, specifically automating process fault analysis. Its ultimate goal was to develop a generalized approach for doing so. Such a generalized approach would allow process fault analyzers to be developed rapidly and inexpensively and to be more easily maintained. This effort led directly to the development of the model-based diagnostic strategy known as the method of minimal evidence (MOME) [1,2]. This inevitably led to the development of a software package known as FALCONEER™ IV [3,4], which automates the MOME algorithm in a fuzzy logic implementation. FALCONEER™ IV

Optimal Automated Process Fault Analysis, First Edition.
Richard J. Fickelscherer and Daniel L. Chester.
© 2013 John Wiley & Sons, Inc. Published 2013 by John Wiley & Sons, Inc.

enables the creation of competent and robust process fault analyzers automatically merely from easily derived engineering models of normal process operation. This converts the solution of the problem of fault analysis directly into solving the much simpler problem of process modeling.

The overall scope and objectives of the original FALCON project are described in greater detail by Lamb et al. [5]. See Rowan [6, 7, 9] and Rowan and Taylor [8] for DuPont's perspective of the FALCON project, and for the Foxboro perspective, see Shirley and Fortin [10] and Shirley [11–13].

B.2 OVERVIEW

The FALCON project investigated developing real-time online knowledge-based systems for actual process control applications, specifically automated process fault analysis. The resulting FALCON system was a knowledge-based system capable of detecting and diagnosing process faults in a commercial adipic acid plant then currently owned and operated by DuPont. In this appendix we briefly describe (1) the operation of DuPont's adipic acid process (2) the FALCON project and its methods and procedures followed for studying and solving the problem of automating process fault analysis, and (3) the four main software modules of the FALCON system developed and used. It concludes by discussing the advantages of using the knowledge-based system paradigm for studying and solving ill-formed real-world problems.

B.3 THE DIAGNOSTIC PHILOSOPHY UNDERLYING THE FALCON SYSTEM

Trade-offs exist in attempting to have a fault analyzer always diagnose every possible process fault situation correctly. As explained in Chapter 3, these trade-offs occur because there exists a spectrum of possible diagnostic sensitivities for the potential process fault situations at each level of diagnostic resolution sought. One extreme of this spectrum represents very sensitive fault analyzers at the highest level of diagnostic resolution. These analyzers sometimes venture diagnoses based on patterns of evidence that are inadequate for the diagnostic discrimination desired. The major drawback of these systems is their tendency to misdiagnose faults either by identifying the wrong fault when a process fault situation exists or by concluding the presence of a fault when none exists. These types of misdiagnoses create very confusing situations for the process operators and could easily lead to inappropriate operator actions. Consequently, fault analyzers performing in such a manner have the

potential to make dangerous process operating situations even more dangerous. DuPont emphasized clearly at the beginning of the FALCON project that
such behavior by the FALCON system would rapidly diminish its credibility
with the process operators and its continued use.

The other extreme of possible fault analyzer sensitivity at the highest
level of diagnostic resolution represents conservative systems that venture
diagnoses based only on definite and complete patterns of evidence. These
systems tend to misdiagnose by concluding that no faults exist when a fault
situation actually does. This type of misdiagnosis may lull the process operators into believing that the process system is operating normally, which in
turn may lead to a serious time delay before appropriate operator actions are
taken. Given their preference, DuPont considered this type of misdiagnosis
to be more tolerable. This is because the process operators are trained to
monitor process operating conditions closely and to follow specific safety
precautions whenever these operating conditions become abnormal. Consequently, the diagnostic knowledge base of the FALCON system developed
was designed to try to ensure that it would misdiagnose only in this manner.
In this capacity, the FALCON system acted as an intelligent operator assistant, giving competent advice if it could, and prudently remaining silent if it
could not.

B.4 TARGET PROCESS SYSTEM

The major motivation for the FALCON project was to study the general issues
involved with developing knowledge-based systems for automated fault analysis. Consequently, selection of the target process system was considered
the most critical factor in the project's eventual outcome. The main characteristics that the target process system should have were therefore defined a
priori. These characteristics are described fully by Lamb et al. [5] and are
summarized here.

The target process system needed to be complicated enough to make the
project's results meaningful. However, it should not be so complicated that
it would be difficult to develop an operable fault analyzer within the time
constraints of the project. The target process system would also need to have
an automated data collection system already in place. A final requirement
was that since the results of this project would eventually be published, the
candidate target process system had to be of relatively low proprietary to
facilitate publication of the results.

DuPont selected the adipic acid process as the target process system for
the FALCON project mainly because it was an established, highly developed,
and widely known technology. Adipic acid is the monomer reactant in the

polymerization of nylon 6,6. Since DuPont was firmly established as one of dominant producers and users of adipic acid worldwide, the threat posed by proprietary information being released inadvertently as a result of the project was minimal. Moreover, a wide variety of serious process fault situations, including explosions, had occurred previously in these plants. Consequently, developing an operable fault analyzer for the adipic acid process system could potentially have some direct safety benefits. It was decided to focus the initial efforts on one particular process subsystem, the adipic acid process's low-temperature converter (LTC) recycle loop. This subsystem is the main reactor of the process and met all of the specified requirements for the project's target process.

A schematic of the process side of the LTC recycle loop is shown in Figure 2.1a. The main process unit in this system is the low-temperature converter (LTC) itself. The LTC is a water-cooled plug flow reactor. Its main function is to remove the heat of reaction evolved from the oxidation via nitric acid (referred to as NAFM) of a mixture of cyclohexanone and cyclohexanol (referred to as TWKA) into adipic acid. The heat of reaction is removed from this highly exothermic reaction by cooling water flowing through the shell side of the LTC. This reaction also produces a relatively large quantity of process offgas. Air is injected into the process stream at the LTC's outlet to dilute the process offgas to reduce the chances that it can ignite. The resulting gas–liquid mixture is separated in a unit called the LTC separator. The gas stream is sent to an absorber where the nitric oxides are recovered and recycled. The process liquid is pumped out of the separator, with the bulk of that liquid being recycled back to the LTC. The residual reactant in the product stream is oxidized in a unit known as the high-temperature converter (HTC). The product stream is then sent to crystallizers, where the adipic acid is removed.

The normal operation of the LTC occurs within a very narrow range of process temperatures. If the process temperature gets too high, the process liquid will degas, causing the process pressure to increase to the point where it will rupture the process equipment. If the process temperature gets too low, the dissolved adipic acid will reach its saturation point and crystallize out of solution. When this occurs, the LTC's tubes become plugged, and only partial oxidation takes place in the LTC. This operating problem, known as *frosting* leads to incomplete reaction within the LTC. The remaining reactants are then converted in the downstream units not designed to remove the heat of reaction. This situation could also cause the process liquid to degas and consequently rupture the process equipment. To minimize the chance of such faults, the process temperature is maintained within the narrow range of temperatures with an elaborate temperature control system. A schematic of the LTC's cooling system is shown in Figure 2.1b.

B.5 THE FALCON SYSTEM

The FALCON system developed consisted of two hardware components and five major software modules. The two hardware components were a DEC MicroVAX II computer with 9 megabytes of RAM memory and an HP 2397A color terminal with touchscreen capability. The five software modules were (1) the inference engine, (2) the human–machine interface, (3) the dynamic simulation model of the adipic acid process, (4) the diagnostic knowledge base, and (5) the process data collection system. The process data collection system was developed by DuPont to interface the FALCON system with the plant's data collection computer system. The other four modules were developed primarily by the University of Delaware. Figure B.1 illustrates how all of these modules are interconnected. The dynamic simulation was the only software module that was not used by the FALCON system when it was operating online in the plant: it was used mainly in the off-line development and verification of the diagnostic knowledge base. Each of the four modules developed by the University of Delaware are described briefly below.

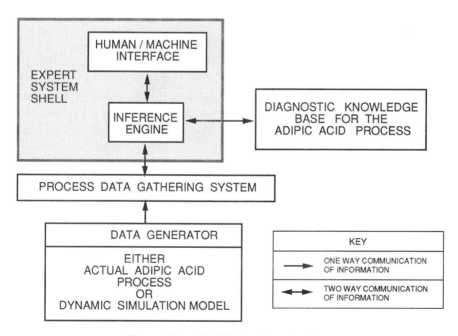

Figure B.1 FALCON system structure.

B.5.1   The Inference Engine

The purpose of the inference engine was to analyze real-time process data with the diagnostic knowledge base for the adipic acid process. The inference engine was written in VAX Common Lisp (over 1600 lines of Lisp code) and ran under a VMS operating system on the DEC MicroVAX II computer. Briefly, the inference engine was designed to (1) accept a time-stamped data vector containing 31 process variables every 15 seconds from the process data collection system, (2) analyze these data with the diagnostic knowledge base, and (3) then report its conclusions to the human–machine interface. Each such inference cycle took approximately 3 CPU seconds to complete. Some additional features were included in the inference engine to check the integrity of the process data obtained from the plant and the communications with the process data collection system and the human–machine interface. Except for these checks, all of the inference engine's conclusions were derived logically from the diagnostic knowledge base. The inferencing procedure used to reach those conclusions is outlined in Figure B.2. The inference engine's logical structure and operation are described in greater detail by Dhurjati et al. [14].

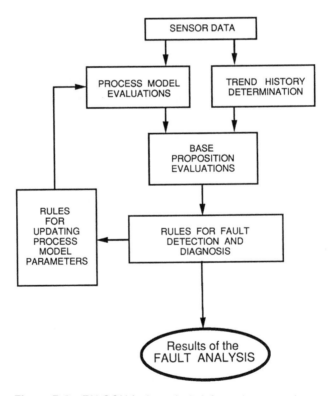

Figure B.2   FALCON fault analysis inferencing procedure.

B.5.2   The Human–Machine Inference

Briefly, the purpose of the human–machine interface was to allow the process operators to understand interactively the reasoning process used by the inference engine to make its diagnoses. For each potential diagnosis the program could also explain why the other possible process fault situations were not plausible. The human–machine interface consisted of the HP 2397A color graphics terminal with touch screen capability and an interactive explanation program written in VAX Fortran-77 (over 4000 lines of Fortran code). The program communicated with the process operators via the touch-screen terminal. In this communication it gave both the inference engine's conclusions and the process information used to reach those conclusions.

The database of the human–machine interface consisted mainly of predetermined explanation text interspliced with variable identifiers. Those identifiers got instantiated with current process information supplied by the inference engine once a fault situation was diagnosed. The predetermined explanation text was organized into a series of interconnected explanation trees, the roots of which were the initial fault announcements made to the process operators. These initial fault announcements were only intended to alert the operators and identify the fault situation present. As such, they did not give supporting evidence for the fault hypothesis tendered, nor did they explain the diagnostic reasoning used to derive that hypothesis. Such information was stored at lower levels in the explanation trees. As one progressed down through those trees, the level of technical detail given by the explanations increased progressively. The program thus allowed the operators to control the level of detail in the explanation they received while reviewing the fault analyzer's conclusions. The program worked in a similar fashion when explaining why other possible process fault situations were not plausible hypotheses. A much more detailed description of the development and operation of the human–machine interface and the lessons learned from its use at the plant were presented by Mooney et al. [15].

B.5.3   The Dynamic Simulation Model

The main purpose of the dynamic simulation model was to act as a plant substitute during the development and testing of the FALCON system (Fickelscherer et al. [16]). This program was developed in an interactive simulation package available at the University of Delaware known as DELSIM. The model was written in Fortran-77. It contained 277 simultaneous ordinary differential equations, over 1050 variables, and over 300 parameters (over 7300 lines of Fortran code). All but seven of the parameters were either the actual physical dimensions of the process equipment or the actual physical properties of various process compounds and mixtures. One of the seven parameters whose value was not equal to its actual process value was the LTC separator level controller proportional band, which had to be detuned for stable operation of the simulation. The other remaining six parameters were estimates of the equivalent length of pipe used in the pressure drop calculations. On average, to get the proper two-phase process and water recirculation flow rates, these had to be set to values five times larger than the values predicted theoretically [17]. No other tuning of the dynamic simulation model was required.

The fourth-order Kutter–Meyerson integration algorithm with variable integration step size was used to solve the model. The average integration step size required was approximately 0.3 second. Solving the model took approximately 3 CPU minutes per minute of simulated process behavior while running under a UNIX operating system on a DEC VAX 11-780 computer. Although this comprehensive dynamic simulation model required major computational efforts to solve, it was definitely felt that the benefits derived from being able to accurately predict the actual process behavior over its entire range of possible operating conditions would be well worth such efforts. It allowed the university personnel to gain a detailed understanding of both normal and abnormal process behavior without spending an inordinate amount of time at the plant and without requiring frequent conversations with busy plant personnel.

The accuracy of the simulation for predicting actual process behavior was verified via a three-step formal verification procedure [1,16]. From this protocol it was determined that the dynamic simulation model was indeed a very accurate representation of the actual process system. The main purpose of the model was to act as a plant substitute during the initial development of the fault analyzer. Process fault situations would be simulated with the dynamic simulation model, and the results would be used to test the performance of the fault analyzer. The dynamic simulation would thus be both a powerful and a highly flexible fault analyzer development tool. Subsequently, the dynamic simulation was used extensively in both the development and verification of the FALCON system's current knowledge base. A total of over 260 process fault situations were simulated and used to test the fault analyzer's performance capabilities.

This need for process fault data was acute. Of all the process fault situations that the fault analyzer was expected to diagnose, actual plant data existed for only about 30% of them. And for all but a few of those fault situations, the actual fault's exact magnitude and rate of occurrence were unknown; they could only be estimated from the process data itself. Furthermore, the available process fault data typically existed for each such fault situation at just one set of the possible process operating conditions.

In the target adipic acid process, an actual process fault situation occurs approximately once a month. Moreover, the ones that do occur reoccur relatively more frequently than the vast majority of the other potential process faults (e.g., LTC frosting). In fact, most of the potential process fault situations actually occur either very infrequently or have never occurred. Consequently, this made obtaining actual process fault data for those faults impossible.

Dynamic simulation models represent the best means of creating these extremely useful fault data. In our case, all of the various possible process fault situations could be simulated at a variety of known magnitudes and rates of occurrence, and over the entire range of process operating conditions. Process fault situations that would be extremely hazardous if they occurred in the plant can be induced quite safely in a dynamic simulation.

Another benefit from creating the dynamic simulation model was derived from consolidating and codifying the various diverse sources of process knowledge that existed. This turned into a recursive procedure for learning about the actual process system's behavior during both normal and abnormal operation. Development of the dynamic simulation model required a highly detailed understanding of the process system's topology and normal operation. This understanding was obtained from (1) a plant visit, (2) process P&I diagrams and operating manuals, (3) Fortran programs for a steady-state process simulation model of the LTC recycle loop and the process system's automatic defrost control algorithm, (4) extensive interviews and correspondance with DuPont engineers; and (5) inspecting actual plant data. Taken together, these diverse information sources contained an enormous wealth of process knowledge that greatly improved our understanding of the adipic acid process system itself and its behavior during both normal and abnormal operation. Elicitation of this knowledge was critical for the eventual success of the FALCON project.

Observing and correcting the behavior of the dynamic simulation model during its development taught us a great deal about the behavior of the target adipic acid process system. It also helped to accelerate us along the learning curve associated with thorough understanding of both normal and abnormal process behavior. Furthermore, this highly detailed, fundamental understanding of process behavior eventually made it possible for us to develop a highly competent fault analyzer. Thus, having first developed the comprehensive dynamic simulation model was a major factor in the eventual success of the FALCON project. The actual process behavior was just too complex for it to be explored adequately by any other means within the time constraints of the project.

Consequently, many invaluable benefits were derived from having available the high-fidelity dynamic simulation model of the adipic acid process during development of the FALCON system. Obviously, however, it is impractical to develop such simulations for every process system for which one wanted to build a process fault analyzer. Many process systems are not understood well enough to enable such high-fidelity dynamic models to be created. Even for those that are understood well enough, some are still so complicated that any benefits derived from using the dynamic simulations would not justify the time and effort required to create them. Moreover, even for those process systems for which accurate dynamic simulations can readily be created, the inordinate amount of time and effort required to exhaustively verify the corresponding fault analyzers' performances using simulated fault data greatly reduces both those fault analyzers' cost-effectiveness and their feasibility. Furthermore, such exhaustive validation strategies would require further that a high-fidelity dynamic simulation be maintained along with the fault analyzer after any process modifications occurred. Such a situation occurred during the FALCON project when the actual adipic acid process was modified. Its associated ramifications to the maintainability of the fault analyzer are described below.

Fortunately, in general, dynamic simulation models of the target process system are not necessary to develop competent and robust fault analyzers. As discussed in detail in Chapter 3, fault analyzers based on the method of minimal evidence (MOME) do not require specific testing of the fault analyzer's underlying diagnostic logic. Rather, competent and robust performance requires only that the engineering models of normal process operation are all *well-formulated*.[1] This major advantage greatly simplifies the creation, verification, and maintenance of these programs, substantially reducing the effort necessary to develop competent and robust fault analyzers by many orders of magnitude over that required by the FALCON project.

B.5.4 The Diagnostic Knowledge Base

The diagnostic knowledge base of the FALCON system was capable of competently diagnosing approximately 160 potential process fault situations, 60 of which could be announced to the process operators. It was created by employing a model-based diagnostic strategy known as the method of minimal evidence [1, 2, 18]. This knowledge base was written in VAX Common Lisp and contained approximately 800 diagnostic rules (over 10,000 lines of Common Lisp code). It was developed over a three-year period in collaboration with DuPont. The knowledge base was tested on over 6000 hours of both actual and simulated plant data to verify its capability to competently diagnose process fault situations in the adipic acid process system.

---

[1]The concept of well-formulated process models is defined in Chapter 2.

B.6 DERIVATION OF THE FALCON DIAGNOSTIC KNOWLEDGE BASE

A brief description of the development effort exerted to create the FALCON system's knowledge base is given here. This includes the knowledge engineering protocols undertaken to study the solution of the fault analysis problem by human experts and how that led to the derivation of the MOME diagnostic strategy to automate process fault analysis [1, 2]. A discussion of the lessons learned from FALCON's performance on plant data during its development was given by Dhurjati et al. [14]. The results of the online plant test were presented by Varrin and Dhurjati [19].

The FALCON system used was the result of a three-year development effort that occurred between December 1984 and January 1988. This development effort mirrored those of other knowledge-based systems reported in the literature.[2] It had four separate stages: (1) rapid prototype development, (2) full system development, (3) full system verification, and (4) a 10-month online plant evaluation. The first two steps were done entirely at the University of Delaware, the third was done jointly at the university and at DuPont's engineering department headquarters (Louviers), and the fourth was performed at the DuPont adipic acid plant (Victoria), with university and Louviers personnel implementing the necessary knowledge base modifications.

B.6.1 First Rapid Prototype of the FALCON System KBS

An initial prototype of the FALCON system was rapidly developed as a proof-of-concept program. Its diagnostic strategy was based on qualitative physics [20]. As discussed in Appendix A, qualitative differential equations used to describe abnormal process operations are called *confluences*. Describing physical systems with confluences correctly is a procedure known as *envisionment*. The resulting fault analyzer was inherently ambiguous, due to the loss of process detail resulting from this envisionment. This diagnostic strategy was quickly abandoned as a methodology for obtaining the desired real-world program performance. It led to a formal search for a more robust diagnostic strategy.

B.6.2 FALCON System Development

The following describes the development effort required to create the FALCON system that was used.

---

[2] A large number of knowledge-based system applications are cited in Section A.9.

B.6.2.1 Better Understanding of the Problem Domain Actual plant data collected during both normal operating conditions and some specific process fault situations became available to the university during 1985. This information source, more than any other, made us appreciate how difficult the task of building a competent real-world fault analyzer was going to be. This program would have to operate correctly in an extremely complex domain, especially during nonnormal process system operation. Being able to predict such operating behavior was essential for proper determination of the required patterns of diagnostic evidence, regardless of the diagnostic strategy utilized, to be employed to correctly classify the current process situation.

As discussed above, the fault analyzer used was developed after the comprehensive dynamic simulation model was completed. This was the university's tool to predict process behavior accurately under all conceivable situations. Once a verified simulation was available, numerous test cases were created to observe the behavior of the process in various operating states, especially in fault modes causing emergency process shutdowns. This added greatly to our understanding of how the process behaved during highly transient situations. Such situations were of interest because the fault analyzer would certainly encounter emergency process shutdowns online in the plant. Consequently, measures would have to be taken to ensure that the fault analyzer would react appropriately.

B.6.2.2 Knowledge Engineering with the DuPont Engineers
While establishing this improved understanding of the problem domain, a search began for a more suitable diagnostic strategy to use in the FALCON system. We met frequently with DuPont engineers and discussed the merits of the various strategies possible. These meetings helped us rule out using methods related to fault tree analysis. Based on their experience, the DuPont engineers believed that in general, developing and maintaining such systems would be too time consuming, too difficult, and too expensive to be practical.

For other practical reasons we also temporarily ruled out trying to incorporate the dynamic simulation into the diagnostic strategy, either as a whole, in parts, or at reduced levels of modeling complexity. The reasons included (1) concerns about the real-time operation of the resulting fault analyzer, (2) problems encountered in our attempts to develop a realistic and efficient diagnostic strategy based on using the dynamic simulation, (3) the large amount of time and effort that would generally be required to develop and verify dynamic simulations accurate enough for this purpose, and (4) the fact that if a dynamic simulation model was used in this manner, it would also have to be maintained as part of the fault analyzer.

One diagnostic strategy discussed that did show promise was quantitative model-based reasoning. Such strategies base their diagnoses on the evaluation of overall conservation equations and process equipment models. However, it appeared from our initial investigation that there were too few useful equations and models derivable for our process system to allow for the unique discrimination of any fault situations. This was due partially to the particular collection of process variables then being monitored and partially to the limited analysis being performed on the models at that time.

B.6.2.3 *Protocol Experiments with the DuPont Engineers* To get past this bottleneck in strategy selection, it was decided that we should more closely investigate the diagnostic strategies used by the actual process engineers. The FALCON project was very fortunate to have access to one particular process engineer, Steve Matusevich, who had over 25 years of experience at Victoria. During the course of the project he was our chief contact at the plant for answers to specific process questions. He also helped to develop and verify the dynamic process simulation, reviewed the human–machine interface explanations, and helped to develop, review, and verify the diagnostic methodology used in the knowledge base. The FALCON project also had access to other DuPont engineers located at Louviers. As they were mostly consultants by profession, they tended to be specialists at solving specific engineering problems rather than being experts for the operation of any particular process system. It was decided to structure the investigation of diagnostic strategies in such a fashion that the effects of these different backgrounds on the engineers' approaches to performing fault analysis could be determined and compared. Two sets of protocol experiments were devised to accomplish this.

The first set of protocol experiments were designed to confront the engineers with simulated process fault situations and then to monitor how they analyzed those data to derive plausible fault hypotheses. During their analysis the engineers would be encouraged to describe what they were thinking about and, if required, to ask for additional process data. After a diagnosis was made, the engineers would be asked to summarize what they believed were the significant reasoning steps they had used to reach their conclusions. If their specific diagnosis was incorrect, the engineers would then be given the opportunity to uncover the mistakes present in their reasoning process. These experiments were called the *mystery fault diagnosis protocols*.

The second set of protocol experiments were designed simply to ask the engineers what response patterns they would expect to observe in the process variables if a particular fault situation was occurring in the plant. These results would be compared to the corresponding results of dynamic simulation runs.

Any discrepancies between the two results would be resolved through discussion. During their analysis the engineers would be encouraged to explain their reasoning and to discriminate between significant and nonsignificant variable responses. This was especially encouraged for the response of process variables that were potentially ambiguous. These experiments were called the *anticipated fault response protocols.*

The mystery fault diagnosis protocols were first performed with Duncan Rowan and Tim Cole, two consultant engineers from Louviers. Both had a familiarity with the adipic acid process system, but they did not consider themselves to be process experts. Without having much experience with the actual process system to guide them in their analysis, they tended to take a systematic and highly structured approach to reach their fault diagnoses. Their approach was based almost exclusively on evidence derived from the evaluation of conservation equations and process equipment models. Although it would sometimes take them a considerable amount of time to reach a conclusion, their diagnoses were generally correct.

Similar protocol experiments and interview meetings were held with Steve Matusevich. These interviews had the dual objective of determining how he performed fault diagnosis and collecting his troubleshooting heuristics. From the protocol experiments we found that he relied heavily on trend analyses, causal reasoning, and troubleshooting heuristics. Consequently, he usually could very quickly recognize and identify process fault situations using just those diagnostic techniques. He would resort to evaluating conservation equations for additional evidence only when he was confronted with very difficult process fault situations: those in which his troubleshooting heuristics did not hold or in which the causal reasoning he used gave poor discrimination between plausible candidate fault hypotheses.

The anticipated fault response protocols were also conducted with Steve Matusevich and Duncan Rowan. From these experiments it was learned that, other than process system topology and an overall understanding of the process system operation, process experts use very little process specific knowledge when predicting process trends: They rely mostly on standard causal reasoning and engineering judgement. With respect to fault situations that could cause a potentially ambiguous response for a given process variable, the expert's judgment was very case-specific. For cases in which the magnitude and/or the propagation rate of a disturbance could not be determined accurately, Steve Matusevich took the position that he would accept any of the possible responses of particular variables as long as they did not violate any possible causality argument. None the less, he was capable of correctly predicting the responses of particular variables in ambiguous situations once approximate estimates of the magnitude and rate of occurrence of the disturbance were known.

From these two sets of protocol experiments we determined that comparable results could be obtained in many process fault situations by using either qualitative causal reasoning or quantitative model-based reasoning. However, in difficult process fault situations, the additional information available in the quantitative models was required to derive unique fault hypotheses. Consequently, it was decided to use quantitative model-based reasoning as the primary diagnostic strategy when developing the fault analyzer.

B.6.2.4 Second Rapid Prototype of the FALCON System KBS

The second prototype of the FALCON system based on quantitative model-based reasoning was completed in February 1986. It was written in Franz Lisp and was run on a DEC VAX 11-780 computer under a UNIX operating system. It could diagnose approximately 50 simulated process faults with a knowledge base that contained approximately 80 diagnostic rules (approximately 1000 lines of Franz Lisp code). The faults being analyzed had been selected by DuPont engineers as a realistic minimum set of possible process problems that DuPont felt the fault analyzer should be able to diagnose. This set became known as the *minimum fault set*. The various process faults included in this set were failure of either process pump; malfunctions in all controllers, control valves, and control variable sensors; restrictions of feed and product flows; and loss of cooling in the LTC. By providing a minimum competency criterion against which its resulting performance would be judged, the minimum fault set gave us a target to focus on during the fault analyzer's development.[3]

To check its soundness, the entire diagnostic methodology utilized by this prototype was reviewed by both DuPont and Foxboro engineers. Meeting their approval, the prototype was then tested with over 100 cases of simulated fault situations. This testing gave encouraging results and also helped to validate the logic of the model-based reasoning employed. Based on these testing results, it was decided to continue developing this prototype into a version that could be field-tested at the plant.

B.6.2.5 FALCON System's Knowledge Base Development

The development of the FALCON system's knowledge base from the prototype's knowledge base took place between March 1986 and mid-September 1986. It caused the knowledge base to increase from approximately 80 diagnostic rules (approximately 1000 lines of Franz Lisp code) to over 800 diagnostic rules (over 10,000 lines of Common Lisp code). This increase in knowledge base size was necessary to handle the complexity inherent in the actual process

---

[3]It set the intended scope of the fault analyzer.

system behavior. The diagnostic strategy employed was improved upon as a result of the lessons learned, with it finally evolving into a more general diagnostic strategy known as MOME. By mid-September 1986, the fault analyzer was delivered to DuPont for an independent verification by Louviers engineers of its performance capabilities.

The enhancements described above made the fault analyzer capable of analyzing the plant for faults over the entire range of process production rates and operating temperatures and during either steady- or unsteady-state operation. The entire FALCON system at that time was rewritten in VAX Common Lisp running under a VMS operating system on a DEC MicroVAX II computer. Once this conversion was completed, a formal demonstration of the FALCON system was given at Louviers to DuPont and Foxboro personnel. At this demonstration, the FALCON system was shown to be capable of correctly diagnosing five plant data files that contained actual process fault situations.

B.6.2.6 Process State Identification Used by the FALCON System   After the formal FALCON system demonstration to DuPont and Foxboro personnel, a comprehensive attempt was begun to formally verify the knowledge base. A major problem that became more apparent during further testing with plant data concerned the process models being used to detect and diagnosis faults. There was a definite need to limit the conditions under which those various models were considered valid representations of the actual process behavior. It was evident that major process upsets such as pump failures and interlock activations invalidated many of the fundamental assumptions used during the development of those models. Consequently, it became necessary that the fault analyzer always determine the current operating state of the process before the diagnostic rules were applied. Such determination would be required to determine which of the specific modeling assumptions were still valid and thus which of the process models would be appropriate in the fault analysis. To do this, process events such as process startups, shutdowns, and interlock activations needed to be monitored explicitly by the FALCON system. The methodology required to perform this type of analysis thoroughly is presented in Appendix C.[4]

Adding this capability to determine the current process state automatically had several consequences. First, it allowed the FALCON system to be turned on when the process was either operating or shut down: The fault analyzer automatically determined which state it was in. It also allowed the FALCON

---

[4]This is the actual pseudocode derived for the original FALCONEER program for monitoring the current operating state of the FMC ESP process.

system to be run continuously, regardless of the plant state transitions that occur. More important, adding this capability logically structured the entire knowledge base according to the patterns of evidence contained within the various diagnostic rules. This reduced the fault analysis of the incoming process data to an ordered, sequential search through the entire set of possible process faults. This search sequence in effect constituted a priority hierarchy between the various possible process fault situations. The logical structure of the fault priority hierarchy within the FALCON system's knowledge base is illustrated in Figure B.3. Discovering the rationale behind this hierarchy represented a major development in the model-based reasoning paradigm used as the primary diagnostic strategy.

These enhancements caused a substantial increase in the size and complexity of the knowledge base. The upgraded version of the FALCON system's knowledge base contained approximately 800 diagnostic rules (over 10,000 lines of Common Lisp code). It was capable of detecting and diagnosing all of the 60 target process fault situations, including the 50 faults specified in the minimum fault set plus an additional 10 sensor faults, plus

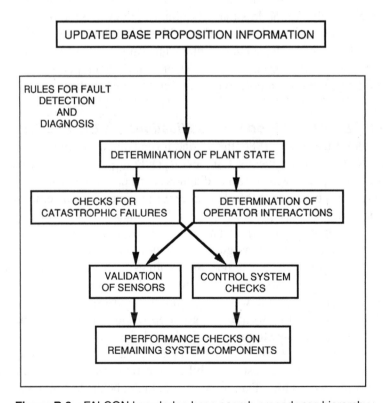

Figure B.3   FALCON knowledge base search precedence hierarchy.

about 100 additional fault situations, most of which were malfunctions in the interlock shutdown system. It was also capable of detecting and diagnosing a few extremely dangerous fault situations that can occur during process startups. This updated version of the FALCON system's knowledge base was tested extensively with plant data throughout the summer of 1986.

B.6.2.7 Lesson Learned Regarding Automating Fault Analysis

Incorporating the few diagnostic rules for the process startup fault situations demonstrated the true potential of rule-based diagnostic programs: It is not necessary to be able to diagnose all possible process fault situations to be able to derive substantial benefits from automated process fault analyzers. We were told by our domain expert, Steve Matusevich, that if any of the rules used to detect and diagnose the process startup faults fired even once, it could potentially eliminate a process accident whose overall cost could pay for the entire FALCON project at least 10 times over (approximately $10 million). This estimate was based on the actual process damage and subsequent downtime caused by one of the aforementioned startup faults that caused an explosion. In a meeting at Louviers between university and DuPont engineers, it took less than 10 minutes to develop the rules required to diagnose those startup fault situations. Moreover, the resulting diagnostic rules very easily could have been implemented independently in a relatively simple FORTRAN program. Clearly, even limited attempts at automating process fault analysis can potentially have enormous benefits.

B.6.2.8 Effects of Process Modifications on the FALCON System

In mid-December 1986, the second formal demonstration of the entire FALCON system was given at Louviers to DuPont personnel. At that time, the FALCON system was shown to diagnose correctly three process fault situations that had occurred in the plant since the previous formal demonstration five months earlier. The progress that had been made in the knowledge base verification at both the university and Louviers was assessed and plans for its completion were discussed in light of a new development at Victoria.

This new development was that the cooling-water system of the target adipic acid process system had been modified two weeks earlier. Consequently, the fault analyzer now had to be modified correspondingly and reverified. It was felt that this would be a good opportunity to examine just how much effort it would take to maintain the knowledge base when such process modifications occurred. Such an understanding was useful to the engineers in further evaluating the costs, flexibility, and time required to maintain a fault analyzer such as this. This certainly would not be the last time such process modification would occur. At Victoria, as in all modern chemical plants, changes occur on a periodic basis as their process systems are improved by their engineers.

The cooling-water system of the adipic acid process had undergone several modifications. One control loop was removed completely and another control signal was rerouted to operate the cooling-water return valve automatically. The effects that these changes would have on the knowledge base were anticipated well in advance of the actual process modification and were outlined in a report submitted to DuPont during the summer of 1986.

One purpose of this report was to request that the cooling water pressure sensor originally scheduled for removal when the modifications occurred be left in place. It had been determined that either this pressure sensor or, alternatively, the addition of a thermocouple to measure the makeup cooling-water temperature would be required to maintain the fault analyzer's current level of diagnostic discrimination. If both measurements were available, it would also be possible to discriminate directly between faults in the two cooling-water thermocouples in all of their possible failure modes. Victoria engineers analyzed the situation and decided to leave the pressure sensor in place.

The other purpose of this report was to request that particular process information be sent to the university once the process modifications had actually occurred. This information was necessary to incorporate the appropriate changes in both the knowledge base and dynamic simulation model. This information included (1) the new tuning constants for the cascaded PID controllers in the cooling-water system, (2) estimates for the expected ranges of fluctuation for the makeup cooling-water temperature and feed pressure, and (3) plant data covering a wide variety of process operating conditions. The plant data were needed primarily to derive a new valve curve correlation for the cooling-water control valve.

Since the changes resulting from the process modification were all localized to the cooling-water system, the knowledge base required only relatively minor modifications. Four potential process fault situations were eliminated by the removal of the one control loop, and only one new potential fault situation had to be analyzed for as a result of the new process configuration. The changes required to the knowledge base were the removal of about 20 diagnostic rules (approximately 160 lines of Common Lisp code) and the addition of about five diagnostic rules (approximately 40 lines of Common Lisp code). It took about 1 hour to determine which rules needed to be removed and to derive the rules that needed to be added. It took about 3 hours to completely implement those changes into the knowledge base. *However, it took three weeks to formally verify that those changes were correct!* And it would have taken much longer had the process side of the adipic acid process been affected by the process modifications. Fortunately, because the modifications had been restricted to the cooling-water system, we were able to determine that the diagnostic rules affected could not interact with the vast majority of the diagnostic rules within the knowledge base.

The reason it took so long to formally reverify the knowledge base was a direct result of the approach we were using to perform it. To adequately test the changes in the knowledge base, we had to simulate 15 separate fault situations. The average simulation run would simulate about 2 hours of process time. This would take between 6 and 8 hours of CPU time to accomplish on one of the university's DEC VAX 11-780 computers. Since it was a time-sharing machine, it would take anywhere between 24 and 36 hours of actual elapsed time, depending on the machine load and the other inefficiencies associated with time sharing, to complete the necessary CPU cycles. Once a simulation was completed, it would take 30 minutes to test the fault analyzer with the simulated fault data, and only about 1 minute to check the results. No changes were required in the knowledge base as a result of this concerted effort: The initial changes made were all correct.

The time delay encountered in the reverification resulted from the fact that we were really performing two verifications at once. The first verification was of the various process modeling equations affected by the plant modification. However, in this situation there were very few modeling equations directly affected. Moreover, for those that were affected, the replacement modeling equations had been verified to be correct when they were derived. The second knowledge base verification being performed was of the patterns of diagnostic evidence used to diagnose the various process fault situations (i.e., the diagnostic rules). It was necessary to verify that these had been derived properly. However, if all of the modeling assumptions associated with the various process modeling equations are known, along with each model's accuracy and its sensitivity to its associated modeling assumptions, it is a very straightforward procedure to derive the proper diagnostic rules. This led us to consider developing ways in which the procedure for deriving diagnostic rules based on MOME could be automated. It eventually led to the creation of FALCONEER™ IV.

B.6.3 The FALCON System's Performance Results

The FALCON system was tested during its development with 260 simulated fault situations and 500 hours of selected process data which contained 65 process operating events (e.g., emergency shutdowns, startups, production changes), including 13 actual process fault situations. During its off-line field test, the FALCON system monitored over 5000 continuous hours of process data in real time. These data included 22 process operating events, eight of which were actual process fault situations. The FALCON system was then tested online for three months. DuPont independently rated its performance at better than 95% correct responses during that test [9].

Although it preformed competently online, maintaining and improving the FALCON system's diagnostic knowledge base (written in Common Lisp based data structures) proved to be impractical for anyone other than the original developer (i.e., the University of Delaware). Being a research project, maintainability was not given as high a priority as was the FALCON system's performance with actual process data.[5] From a research viewpoint, generalizing the underlying logic of this model-based diagnostic strategy was paramount, allowing future such development project activities to be as streamlined as possible. This effort led directly to the formulation of MOME.[6]

B.7   THE IDEAL FALCON SYSTEM

It became apparent during the FALCON project that because of the nature of the problem being solved, the fault analyzer's development could never truly be finished: As discussed, incremental improvements would continue to be required periodically as the process system and its operation evolved. Consequently, we decided to characterize the performance of the ideal FALCON system so that we could gauge how close we were to achieving such ideal performance. The following defines our ideal performance criteria for automated process fault analyzers:

1. The fault analyzer will be able to correctly classify all possible significant single process operating events (i.e., it will not diagnose any of those events incorrectly, nor will it diagnose a fault situation when none exists).
2. The fault analyzer will be able to function over all possible operating rates and at any of the normal operating conditions while performing criterion 1.

---

[5]This is quite common to such development efforts. As Stock [21] states: "The ad hoc procedure of architecture, design, and implementation associated with incremental development leads to an unwieldy, unstructured application that is extremely difficult to maintain, modify, integrate, and extend as new requirements become necessary."

[6]Jackson [22] states: "This approach is very much in keeping with the tradition of 'standing back' from some working program to look at it from a higher level of abstraction, in order to see what has actually been learned from its implementation, and then performing a rational reconstruction which both extends the power of the original program and achieves its ends in a more principled way. There is much to be said for this as a research strategy in expert systems and artificial intelligence generally."

3. The fault analyzer will correctly monitor both operator- and control system-initiated changes and will also monitor correctly all normal transients in the process state resulting from those changes.
4. The fault analyzer will correctly monitor for faults prior to and during any emergency process interlock activations.
5. The fault analyzer will be able to be turned on during any possible process state, determine that state automatically, and then monitor for all state transitions possible from that state.
6. The fault analyzer will be able to monitor the incoming data and determine whether those data are unique, in the correct format and within the proper limits, and then adjust its analysis accordingly.

To summarize, these criteria state that the ideal fault analyzer will operate properly under all circumstances and will correctly diagnose all single significant process faults that can occur during the production phase of process operations. Doing so mirrors an ideal concept of intelligence called *rationality*; that is, a system is rational if it does the right thing [23]. When this list of criteria was compiled, it was believed that the FALCON system was actually very close to being ideal. Minor problems still existed that would have to be dealt with in time, but all in all, the basic development and verification of the fault analyzer had been completed. The experience and lessons learned from actually creating a competent real-world fault analyzer led directly to a conceptual algorithm capable enough to be a robust solution to the general automated process fault analysis problem.

B.8 USE OF THE KNOWLEDGE-BASED SYSTEM PARADIGM IN PROBLEM SOLVING

The true power of a knowledge-based systems' approach to problem solving is that it can be used to quickly automate solutions to poorly understood problems. This approach allows one to capture, organize, and directly utilize the diverse forms of domain and procedural knowledge required to solve those problems. The resulting solutions separate the domain-specific (i.e., application-specific) knowledge from the underlying procedures or algorithms (i.e., generalized methods for solving problems) followed to obtain those solutions. Separating the two types of knowledge in this manner directly facilitates the study of poorly understood problems, leading eventually to generalized solutions for those problems. Once obtained, these generalized solutions can be implemented in a variety of ways, possibly even in conventional computer hardware and software.

This was the main use of the knowledge-based system paradigm throughout the FALCON project. As described above, this approach to problem solving was therefore not an end in itself, but rather, was a means to an end. It allowed for rapid prototyping of potential solutions to the problem of automated process fault analysis, examination of the limitations of those solutions, and incremental evolution to more robust solutions. Using this paradigm eventually led to the derivation of the MOME quantitative model-based diagnostic strategy.

REFERENCES

1. Fickelscherer, R. J., *Automated Process Fault Analysis*, Ph.D. dissertation, Department of Chemical Engineering, University of Delaware, Newark, DE, 1990.
2. Fickelscherer, R. J., "A Generalized Approach to Model-Based Process Fault Analysis," in *Foundations of Computer-Aided Process Operations II*, ed. by D. W. T. Rippin, J. C. Hale, and J. F. Davis, CACHE, Inc., Austin, TX, 1994, pp. 451–456.
3. Chester, D. L., L. Daniels, R. J. Fickelscherer, and D. H. Lenz, U.S. Patent 7,451,003, "Method and System of Monitoring, Sensor Validation and Predictive Fault Analysis," 2008.
4. Fickelscherer, R. J., Lenz, D. H., and Chester, D. L., "Fuzzy Logic Clarifies Operations," *InTech*, October 2005, pp. 53–57.
5. Lamb, D. E., D. L. Chester, P. S. Dhurjati, and J. C. Hale, "An Academic/Industry Project to Develop an Expert System for Chemical Process Fault Detection," Paper 70c, presented at the *AIChE Annual Meeting*, Chicago, November 1985.
6. Rowan, D. A., "Chemical Plant Fault Diagnosis Using Expert Systems Technology: A Case Study," in *IFAC Kyoto Workshop on Fault Detection and Safety in Chemical Plants*, 1986, pp. 81–87.
7. Rowan, D. A., "Using an Expert System for Fault Diagnosis," *Control Engineering*, 1987, pp. 160–164.
8. Rowan, D. A., and R. J. Taylor, "On-Line Fault Diagnosis: FALCON Project," in *Artificial Intelligence Handbook*, Vol. 2, Instrument Society of America, Research Triangle Park, NC, 1989, pp. 379–399.
9. Rowan, D. A., "Beyond FALCON: Industrial Applications of Knowledge-Based Systems," in *Proceedings of the International Federation of Automatic Control Symposium*, Newark, DE, ed. by P. S. Dhurjati, 1992, pp. 215–217.
10. Shirley, R. S., and D. A. Fortin, "Status Report: An Expert System to Aid Process Control," in *Proceedings of ISA/85*, 1985, pp. 1463–1470.
11. Shirley, R. S., "Status Report 2: An Expert System to Aid Process Control," in *Proceedings of the TAPPI Engineering Conference*, 1986, pp. 425–430.

12. Shirley, R. S., "Status Report 3 with Lessons: An Expert System to Aid Process Control," in *Proceedings of the Annual Pulp and Paper Industry Technical Conference*, 1987, pp. 132–136.

13. Shirley, R. S., "Some Lessons Learned Using Expert Systems for Process Control," in *Proceedings of the American Control Conference*, Vol. 2, 1987, pp. 1342–1346.

14. Dhurjati, P. S., D. E. Lamb, and D. L. Chester, "Experience in the Development of an Expert System for Fault Diagnosis in a Commercial Scale Chemical Process," in *Proceedings of the First International Conference on Foundations of Computer Aided Process Operations*, ed. by G. V. Reklaitis and H. D. Spriggs, Elsevier Science, New York, 1987, pp. 589–626.

15. Mooney, D. J., D. L. Chester, D. E. Lamb, and P. S. Dhurjati, "Design and Operation of the FALCON Interface", in *Proceedings of ISA/88*, 1988, pp. 747–758.

16. Fickelscherer, R. J., P. S. Dhurjati, D. E. Lamb, and D. L. Chester, "Role of Dynamic Simulation in the Construction of Expert Systems for Process Fault Diagnosis," Paper 51d, presented at the *AIChE Spring National Meeting*, New Orleans, LA, 1986.

17. Olujic. Z., "Compute Friction Factors Fast for Flow in Pipes," *Chemical Engineering*, Vol. 88, No. 25, 1981, pp. 91–93.

18. Fickelscherer, R. J., P. S. Dhurjati, D. E. Lamb, and D. L. Chester, "The FALCON Project: Application of an Expert System to Process Fault Diagnosis," Paper 82a, presented at the *AIChE Spring National Meeting*, Houston, TX, 1987.

19. Varrin, R. D., Jr., and P. S. Dhurjati, "Implementation of an Expert System for On-Line Fault Diagnosis in a Commercial Scale Chemical Process," Paper 21b, presented at the *AIChE Spring National Meeting*, New Orleans, LA, 1988.

20. Chester, D. L., D. E. Lamb, and P. S. Dhurjati, "An Expert System Approach to On-Line Alarm Analysis in Power and Process Plants," *Computers in Engineering ASME*, Vol. 1, 1983, pp. 345–351.

21. Stock, M., *AI in Process Control*, McGraw-Hill, New York, 1989, p. 140.

22. Jackson, P., *Introduction to Expert Systems*, Addison-Wesley, Reading, MA, 1986, p. 114.

23. Russell, S. J., and P. Norvig, *Artificial Intelligence: A Modern Approach*, Prentice Hall, Upper Saddle River, NJ, 1995, pp. 826–830.

C

PROCESS STATE TRANSITION LOGIC USED BY THE ORIGINAL FALCONEER KBS

C.1 INTRODUCTION

In this appendix we describe in detail how the original (i.e., hand-compiled) version of the FMC Tonawanda ESP knowledge-based system (KBS; a.k.a. FALCONEER) determined the current operating state of the electrolytic sodium persulfate process. For the same reasons that doing so was required in the original FALCON system, this was again essential for the KBS to perform proper analysis of the process sensor data collected. The program was able to be turned on at any time and would automatically determine its current process state and monitor for all state transitions possible from that state. This allowed it to minimize nuisance alarms while doing its intelligent supervision of that process's operation.

C.2 POSSIBLE PROCESS OPERATING STATES

There are seven possible process operating states in FMC's ESP process:

0. The process has been shut down previously and is awaiting the next startup.

Optimal Automated Process Fault Analysis, First Edition.
Richard J. Fickelscherer and Daniel L. Chester.
© 2013 John Wiley & Sons, Inc. Published 2013 by John Wiley & Sons, Inc.

1. The process is being started up: current to cells is on but cell products are being recycled to the neutral tank rather than being sent to the crystallizer. Once this switch to the crystallizer is accomplished, the process is considered to be in production mode.
2. The process is in production mode and is running within all *standard operating conditions* (SOCs) of its key sensor points (i.e., relevant process variables). These points are those with associated high and/or low interlock activation limits.
3. The process is in production mode but is not running within all SOCs of its key sensor points.
4. The process is in production mode but is rapidly approaching one or more interlock regions of operation.
5. One or more process interlocks should have occurred, but the process is still in production mode.
6. The process is just shutting down (current to the cells has just been shut off).

Possible ESP Process State Transitions

It is possible to go from state 0 to state 1, 2, 3, 4, or 5.
It is possible to go from state 1 to state 2, 3, 4, 5, or 6.
It is possible to go from state 2 to state 1, 3, 4, 5, or 6.
It is possible to go from state 3 to state 1, 2, 4, 5, or 6.
It is possible to go from state 4 to state 1, 2, 3, 5, or 6.
It is possible to go from state 5 to state 1, 2, 3, 4, or 6.
It is possible to go from state 6 to state 0, 1, 2, 3, 4, or 5.

These possible ESP process state transitions indicate that almost any other state can be reached from any starting state. This occurs because these transitions can happen faster than the monitoring data sampling rate of the FALCONEER KBS (currently the analysis sampling interval is 1 minute). It can, consequently, miss the commencement of significant operating events leading to abnormal operating behavior. Thus, the KBS must be prepared to encounter any state at any time. By definition, transversal of the possible process states from state 1 to state 6 will be defined as a completed production campaign.

C.3   SIGNIFICANCE OF PROCESS STATE IDENTIFICATION AND TRANSITION DETECTION

The program must determine which state the process is in so that only meaningful data are analyzed and validated. The fault analyzer will be active during process states 2, 3, and 4 for determining all possible assumption variable deviations and will perform interlock malfunction analysis during process state 5. The KBS is in waiting mode (state 0) after shutdown (state 6) occurs and will monitor the process passively to determine when the next startup (state 1) will be complete. It will then become active again automatically.

C.4   METHODOLOGY FOR DETERMINING PROCESS STATE IDENTIFICATION

The process state that the process is currently in when the program is turned on is inferred from the present-value states and predicted next-value states of the various key sensor variables. A process sensor variable is considered a key if it can cause interlock shutdown. The present-value and predicted next-value states of these measurements are determined as described below.

C.4.1   Present-Value States of All Key Sensor Data

Each key sensor point will be used to populate the following sensor record. Each field in this record will be updated if required whenever a new vector of data is obtained by the *process historical data* (PHD) server connected to the *distributed control system* (DCS).

(1) PV = present value

(2) PVS = present-value state

(3) PPV = previous present value

(4) PNV = predicted next value

(5) PNVS = predicted next-value state

(6) MaxDCS = maximum DCS control system value

(7) HIntlk = high interlock limit

(8) MaxSOC = maximum standard operating condition

(9) MinSOC = minimum standard operating condition

(10) LIntlk = low interlock limit

(11) MinDCS = minimum DCS control system value

Table C.1   Determination of Present-Value State

Possible Present Values (PV)				PV State (PVS)	PVS ID
MaxDCS	< PV		→	Bad value high	3
HIntlk	≤ PV	≤ MaxDCS	→	High interlock expected	2
MaxSOC	< PV	< HIntlk	→	Outside maximum SOC	1
MinSOC	≤ PV	≤ Max SOC	→	Within SOC	0
LIntlk	< PV	< MinSOC	→	Outside minimum SOC	−1
MinDCS	≤ PV	≤ LIntlk	→	Low interlock expected	−2
	PV	< MinDCS	→	Bad value low	−3

At any given time, each key sensor can be in one and only one state of the seven possible states for its *present-value state* (PVS). These states are determined as shown in Table C.1.

C.4.2   Predicted Next-Value States of All Key Sensor Data

The *predicted next value* (PNV) for a given key sensor PV is calculated by a simple linear extrapolation of that PV and the last previously measured PV, referred to as the *previous present value* (PPV). The formula for calculating the PNV is

$$PNV = 2PV - PPV \qquad (C.1)$$

The PNV is useful for anticipating process state transitions to state 4. This is determined as shown in Table C.2.

Anticipating interlock activations before they happen will help eliminate spurious or incorrect KBS diagnoses from occurring. After an interlock trip occurs, the various primary and secondary model residuals are not evaluated to avoid a GIGO (i.e., garbage in, garbage out) situation with those residuals. In these situations, the KBS would passively monitor the process until it could determine that state 1 (startup) is completed for the next production campaign and the KBS data buffers are flushed. It would then automatically begin to analyze process data again.

Table C.2   Determination of Predicted Next Value State

Possible Predicted Next Values (PNV)				PNV State (PNVS)	PNVS ID
HInlk	≤ PNV		→	Expect high interlock soon	1
LInlk	< PNV	< HInlk	→	Interlock not expected soon	0
	PNV	≤ HInlk	→	Expect low interlock soon	−1

C.5   PROCESS STATE IDENTIFICATION AND TRANSITION LOGIC PSEUDOCODE

The following describes the logic for determining the current process state and state transitions for the FMC ESP process. This logic is written in object-oriented programming pseudocode, where objects have both specific attributes and methods for determining values for those attributes. The logic described assumes that all process variable measurements used by the KBS arrive periodically as a complete time-stamped vector. The KBS then uses those values and any past results that it requires from previous KBS analyses to analyze this current vector and to display its results. This analysis must be completed before the next vector of process variable measurements arrives to allow the KBS to run in *real time*. Currently, the FALCONEER KBS uses a period of 1 minute as the update frequency for data analyzed by the KBS.

C.5.1   Attributes of the Current Data Vector

Attribute	Data Type	Description
(1) first_data_received	Boolean	This is the first vector of process sensor data to be analyzed by the KBS for this production campaign.
(2) KBS_initialized	Boolean	KBS data buffers have been flushed with legitimate data (i.e., those collected during actual operating conditions).
(3) FALCONEER _monitoring_mode	Integer	**0**: The KBS is passively monitoring process until the next startup is complete. **1**: The KBS is actively performing sensor validation and proactive fault analysis of all possible modeling assumption deviations being monitored directly. **2**: The KBS is actively performing interlock failure analysis for all interlocks within its realm of process system operation. (*This analysis is currently outside the scope of the FALCONEER KBS, but the KBS's current design allows for it to be added directly.*)
(4) Current_to_Cells_is_on	Boolean	Current to the cells is now on.
(5) Product_being_sent_to _Crystallizer	Boolean	The cell product is not being recycled but actually is sent to the crystallizer.

(6) all_process_variables _within_SOC	Boolean	All key process sensors are within their standard operating condition (SOC) limits.
(7) Sensor_PNVS(I)	Integer array	The predicted next-value state for each key process sensor being monitored; it is assigned one of three possible state values, two which indicate that an interlock low (-1) or high (1) is expected from it soon.
(8) interlock_expected _soon	Boolean	An interlock trip is anticipated within the next time step interval.
(9) Sensor_PVS(I)	Integer array	The present-value state for each key process sensor being monitored; it is assigned one of seven possible state values, four which indicate that an interlock low (-2 or -3) or high (2 or 3) is now expected.
(10) interlock_expected _now	Boolean	An interlock trip is anticipated now.
(11) startup_commenced	Boolean	Current to the cells is on but product is still being recycled to the neutral tank.
(12) startup_completed	Boolean	The process is in production mode.
(13) shutdown _commenced	Boolean	Current to the cells has been shut off.
(14) current_process_state	Integer	**0**: The process state is unknown. (The KBS is idle awaiting the next process startup.)
		1: Process startup is occurring. (Current to the cells is on, but the product is being recycled.)
		2: The process is in production and is within all SOCs.
		3: The process is in production but is not within all SOCs.
		4: The process is in production but is rapidly approaching one or more interlock(s).
		5: Process interlock(s) should be occurring.
		6: The process has just shut down.

C.5.2 Method Applied to Each Data Vector

Before any data vectors have been received, attribute `first_data_received` is set to FALSE. Then, as each data vector is received, the sequence of statements presented below are executed.

C.5.2.1 *determine_if_first_data_received* Initialize all the various Booleans after the KBS receives its first vector of sensor data collected for either the current or next (if the process is in state 6) production campaign.

```
IF (first_data_received EQ FALSE) THEN
          startup_commenced       = FALSE
          startup_completed       = FALSE
          shutdown_commenced      = FALSE
          KBS_initialized         = FALSE
          first_data_received     = TRUE
END IF
```

C.5.2.2 *determine_if_startup_commenced* Determine if the current is on to cells.

```
IF ((IIC8000R.PVS > -2 ) AND (IIC8000R.PVS < 3 )) THEN
          Current_to_Cells_is_on = TRUE
ELSE
          Current_to_Cells_is_on = FALSE
END IF
IF (( Current_to_Cells_is_on EQ TRUE) AND
          (startup_commenced EQ FALSE)) THEN
          startup_commenced = TRUE
END IF
```

C.5.2.3 *determine_if_startup_completed* Determine if the process is in production mode.

```
IF (XCV6272.PV EQ NORMAL) THEN
          Product_being_sent_to_Crystallizer  = TRUE
ELSE
          Product_being_sent_to_Crystallizer  = FALSE
END IF
IF ((startup_commenced EQ TRUE) AND
          (Product_being_sent_to_Crystallizer EQ TRUE) AND
          (startup_completed EQ FALSE)) THEN
          startup_completed = TRUE
END IF
IF ((startup_completed EQ TRUE) AND
          (Product_being_sent_to_Crystallizer EQ FALSE))
THEN
          startup_completed = FALSE
END IF
```

C.5.2.4 determine_if_shutdown_commenced Determine if the process is being shut down.

```
IF ((Current_to_Cells_is_on EQ FALSE) AND
        (startup_commenced EQ TRUE) AND
        (shutdown_commenced EQ FALSE)) THEN
            shutdown_commenced = TRUE
END IF
```

C.5.2.5 determine_current_process_state Assigns one of seven possible values to the current process state.

Determine whether all process variables are within the SOCs.

```
all_process_variables_within_SOC = TRUE
FOR EACH I FROM 1 TO Number_of_Key_Sensor_Variables DO
        IF (Sensor.PVS(I) NE 0 ) THEN
                all_process_variables_within_SOC = FALSE
        END IF
```

Determine whether interlock is expected soon.

```
interlock_expected_soon = FALSE
FOR EACH I FROM 1 TO Number_of_Key_Sensor_Variables DO
        IF (Sensor.PNVS(I) EQ (-1 OR 1)) THEN
                interlock_expected_soon = TRUE
END IF
```

Determine whether interlock is expected now.

```
interlock_expected_now = FALSE
FOR EACH I FROM 1 TO Number_of_Key_Sensor_Variables DO
        IF (Sensor.PVS(I) EQ (-2 OR -3 OR 2 OR 3)) THEN
                interlock_expected_now = TRUE
        END IF
```

Assign the current process state. If the current operating state does not match any of the first six possible states (1, 2, 3, 4, 5, or 6), it is assigned the unknown state (0).

State (1): The process is being started up (the KBS will not allow autopilot mode and it is not performing fault analysis).

```
IF ((startup_commenced EQ TRUE) AND
        (startup_completed EQ FALSE)) THEN
        current_process_state = 1
```

State (2): The process is running and all key sensor variables are within SOCs (autopilot mode allowed once the KBS has been initialized).

```
ELSE IF ((startup_completed EQ TRUE) AND
         (all_process_variables_within_SOC EQ TRUE) AND
         (interlock_expected_soon EQ FALSE)) THEN
         current_process_state = 2
```

State (3): The process is running but not all key sensor variables are within SOCs (ask the operators if they still want autopilot mode after the KBS is initialized).

```
ELSE IF ((startup_completed EQ TRUE) AND
         (all_process_variables_within_SOC EQ FALSE) AND
         (interlock_expected_soon EQ FALSE) AND
         (interlock_expected_now EQ FALSE)) THEN
         current_process_state = 3
```

State (4): The process is running and is quickly approaching interlock shutdown (drop out of autopilot mode and alert the operator that immediate action is necessary).

```
ELSE IF ((startup_completed EQ TRUE) AND
         (interlock_expected_soon EQ TRUE) AND
         (interlock_expected_now EQ FALSE)) THEN
         current_process_state = 4
```

State (5): The process is just about to interlock or there has been an interlock activation failure. Actively perform interlock failure analysis (not currently within the original FALCONEER KBS's intended scope).

```
ELSE IF ((startup_completed EQ TRUE) AND
         (interlock_expected_now EQ TRUE) AND
         (Current_to_Cells_is_on EQ TRUE)) THEN
         current_process_state = 5
```

State (6): The process has just been shut down; reset all the state transition Booleans once process shutdown occurs so that the KBS will monitor properly for the next process startup.

```
ELSE IF (shutdown_commenced EQ TRUE) THEN
         current_process_state = 6
         first_data_received = FALSE
```

State (0): The process has been shut down and its restart status is unknown. The KBS monitors data passively until the start of the next production campaign occurs.

```
ELSE
        current_process_state = 0
END IF
```

C.5.2.6 determine_if_KBS_initialized Sets a Boolean indicating that the KBS is actively performing its analysis.

```
IF ((current_process_state EQ (2 OR 3 OR 4 OR 5)) AND
        (KBS_initialized EQ FALSE)) THEN
        KBS_initialized = TRUE
END IF
```

C.5.2.7 determine_FALCONEER_monitoring_mode Sets the monitoring mode of the KBS on the current process sensor data.

```
IF ((KBS_initialized EQ TRUE) AND
        (current_process_state EQ (2 OR 3 OR 4))) THEN
        FALCONEER_monitoring_mode = 1
ELSE IF ((KBS_initialized EQ TRUE) AND
        (current_process_state EQ 5)) THEN
        FALCONEER_monitoring_mode = 2
ELSE
        FALCONEER_monitoring_mode = 0
        KBS_initialized = FALSE
END IF
```

C.6 SUMMARY

The temporal logic described above as object-oriented program pseudocode has been demonstrated to be effective in the original FALCONEER KBS for determining the current KBS context. Doing so allows the KBS to monitor process operations continuously and perform its sensor validation and proactive fault analysis (SV&PFA) on all legitimate process data, shutting off its analysis when it is not appropriate to perform it. This automatically eliminates GIGO situations.

D

FALCONEER™ IV REAL-TIME SUITE PROCESS PERFORMANCE SOLUTIONS DEMOS

D.1 FALCONEER™ IV DEMOS OVERVIEW

There are two FALCONEER™ IV process application demonstrations and associated process fault data downloadable from our website to show some of FALCONEER™ IV functions and capabilities: specifically, (1) a wastewater treatment process and (2) a pulp and paper process. To run a demo, make sure that FALCONEER™ IV is installed. Browse to the demos folder at our website and follow the installation instructions recommended:

```
http://falconeertech.com/8843/228823.html
```

D.2 FALCONEER™ IV DEMOS

The following two demonstrations are currently available.

D.2.1 Wastewater Treatment Process Demo

This FALCONEER™ IV demonstration is a simple tank with several inputs and reactions. It represents a wastewater treatment basin with aeration for

---

Optimal Automated Process Fault Analysis, First Edition.
Richard J. Fickelscherer and Daniel L. Chester.
© 2013 John Wiley & Sons, Inc. Published 2013 by John Wiley & Sons, Inc.

biological and chemical treatment. Although simple, the demonstration shows many of FALCONEER™ IV's real-time capabilities:

1. Monitor and detect out-of-control conditions using the virtual statistical process control (virtual SPC, the rightmost LED column on the alarm screen).
2. Monitor and alarm on key performance conditions, such as excess energy cost.
3. Create and monitor a performance equation (in this case, calculated dissolved oxygen) with a virtual or soft sensor;
4. Monitor, detect, and alert on instrumentation problems using sensor validation and proactive fault analysis (SV&PFA, the middle LED column on the alarm screen);
5. Continuously validate the reliability of instrumentation and equipment.

The demo begins in normal steady-state operation. Wastewater influent and effluent flow is 16,000,000 gal/day with enough aeration at 11,000 SCFM to keep the solids suspended and the dissolved oxygen (DO) concentration in the basin at around 2.0 ppm. There are two key operational and cost performance conditions being calculated and monitored. The first, "% overaeration of the basin," directly affects the second, "excess electrical costs due to over-aeration." The goal is to maintain a DO of around 2.0 ppm to meet effluent permit requirements while operating at the lowest energy cost. To ensure the reliability of this important instrument, a redundant virtual or soft DO meter is created, calculating DO in realtime for comparison to the measured DO. All instrumentation is validated, as indicated by the green LEDs in the middle column. All measured conditions are in control, as indicated by the green LEDs in the right column. None of the instrumentation is used for state ID, so the LEDs in the left column are all grayed out.

Starting at 1:19 A.M., airflow is increased by about 5%, triggering the virtual SPC alarm for the airflow. In this demo, the airflow is an uncontrolled variable and the control limits adjust accordingly, so the virtual SPC alarm changes from red to yellow and finally, to green at the new steady state. The DO is increased to about 2.4 ppm, slightly higher than needed. The two key performance conditions, being monitored in real time using virtual SPC, trigger warning yellow alarms, indicating that continued operation at this condition will cost about $9,000 a year more than necessary.

Starting at 1:37 A.M., airflow is increased by 37%, again triggering the virtual SPC alarm for the airflow that goes away as the control limits adjust. The two performance conditions are exceedingly high and red virtual SPC alarms are triggered, indicating that continued operation at this condition will cost about $69,000 per year more than necessary. The DO is increased to greater than 4.0 ppm, which for this demo exceeds the upper precontrol limit (user adjustable) to further reinforce that this condition is significantly higher than needed. Both the DO measured and that calculated confirm this abnormal condition. The condition is returned to normal at 1:58 A.M. and the SPC alarms go away.

Starting at 2:10 A.M., the DO meter becomes faulty, incurring a drift or offset of −0.5 ppm. For this demo it is assumed that if the DO falls below 1.8 ppm, continued operation may result in permit violations of the effluent. However, a red virtual SPC alarm is first triggered on the actual DO meter, but not the virtual DO meter, providing early warning of a problem with the instrument. Then the SV&PFA diagnosis provides a yellow warning followed by a red alarm, indicating that there is a fault or failure or abnormal process condition associated with the actual DO meter, and attention is required by the operator and/or maintenance. The calculated or soft DO meter remains validated and useful for continued operational control. The faulty meter is assumed to be fixed at 2:30 A.M. and the alarms go away.

Finally, starting at 2:42 A.M., the effluent flowmeter becomes faulty, incurring a drift or offset of + 1,000,000 gal/day. Again, the red virtual SPC alarm is triggered, followed by a yellow SV&PFA warning, then a red SV&PFA alarm. The remaining instrumentation remains correct and validated by the real-time validation provided by the SV&PFA. The flowmeter is assumed to be fixed at 3:02 A.M. and the alarms go away. When this final effluent flow meter alarm ends, the FALCONEER™ IV is simply stopped and then restarted to replay the demo and explore the other screens and reports.

D.2.2   Pulp and Paper Stock Chest Demo

The following demo describes the behavior of FALCONEER™ IV when monitoring a single stock chest unit operation. Stock chests are common systems to all pulp mills with papermaking machines. A stock chest is basically a continuous mixing tank connected to a mixing tee where dilution water is added to a pulp mixture to control the pulp concentration or consistency of the stream being sent to the paper machine. All of the screens described below are accessible from the Windows menu item on the FALCONEER™ IV client screen.

The following measurements are available for continuous monitoring by FALCONEER™ IV:

1. Inlet consistency (fraction)
2. Inlet flow rate (gpm)

3. Inlet temperature (°F)
4. Stock chest level (%)
5. Stock chest temperature (°F)
6. Dilution water header pressure (psig)
7. Dilution water control valve opening (%)
8. Outlet consistency (fraction)
9. Outlet flow rate (gpm)
10. Outlet temperature (°F)

With these measurements it is possible to calculate the dilution water flow rate from the dilution water header pressure and control valve opening, and the stock chest consistency from this dilution water flow rate and the outlet flow rate and consistency. The temperature of the dilution water is assumed to be known and constant at a value of 55 °F.

With the 10 variables listed above it is possible to derive five linearly independent process models:

1. Overall dynamic mass balance
2. Overall dynamic energy balance
3. Overall dynamic pulp balance
4. Mixing tee energy balance
5. Upstream steady-state inlet consistency balance

With these five models it is possible to diagnose faults occurring in each of the measurements or parameters listed above as well as to detect leaks in the stock chest, dilution water feed line, or stock chest outlet feed line. FALCONEER IV automatically and exhaustively combines these five linearly independent models to create 13 linearly dependent models which help improve its diagnostic resolution for the various potential faults [i.e., most unique resolution between all possible occurring faults to being only the most plausible (i.e., highest associated certainty) fault hypotheses].

The program continuously monitors the 10 measured variables, evaluates all the models to compute model residuals (these should all evaluate to 0.0 when there are no faults occurring), and converts those residuals into certainties between -1.0 (the corresponding residual is definitely low) and $+1.0$ (the corresponding residual is definitely high). It then combines the relevant model certainties according to our proprietary fuzzy logic diagnostic rule to derive all plausible fault hypotheses. These are presented to the user as both alarms on the middle column of the alarm screen and as a program trace of the fault analysis performed to derive these alarms on the faults and models screen. The current values of all the measurements and other constant or calculated parameters used by FALCONEER™ IV in its fault analysis are displayed on the current values screen.

In addition to fault analysis, FALCONEER™ IV continuously performs virtual statistical process control as desired on all measured and calculated variables. In the stock chest demonstration, four measured variables are monitored by virtual SPC. These are:

1. Inlet consistency
2. Outlet consistency
3. Dilution water pressure
4. Outlet flow

The current statistics of these four variables are displayed on the virtual SPC screen. When alarming, they are also shown on the rightmost column of the alarm screen.

The stock chest data file given here demonstrates FALCONEER™ IV's performance when there is a bias of $+0.01$ on the inlet consistency meter (i.e., its measured value becomes 0.055 while its actual value remains 0.045). This step change occurs at 2:40 A.M. FALCONEER™ IV immediately catches this fault and gives two corresponding alarms on the alarm screen that the measurement has failed high and that the measurement is out of control high. These alarms remain until the virtual SPC condition eventually goes away after the virtual SPC calculation re-zeros with the new measurement. At 3:07 A.M. the bias is slowly eliminated in a drift that lasts until 3:26 A.M., at which time the fault announcement also goes away. As described above, details of the various computations being performed are shown in the associated screens reachable from the Windows menu item.

When the inlet consistency meter alarm finally ends, FALCONEER™ IV is simply stopped and then restarted to replay the demo and explore the other screens and reports.

INDEX

Optimal Automated Process Fault Analysis, First Edition.
Richard J. Fickelscherer and Daniel L. Chester.
© 2013 John Wiley & Sons, Inc. Published 2013 by John Wiley & Sons, Inc.